VIROLOGY RESEARCH PROGRESS

PARVOVIRUS B19 AND HEMATOLOGICAL DISORDERS IN CHILDREN

VIROLOGY RESEARCH PROGRESS

Herpesviridae: Viral Structure, Life Cycle and Infections
Toma R. Gluckman (Editor)
2009. ISBN: 978-1-60692-947-6
2009. ISBN: 978-1-60876-921-6 (E-book)

Insect Viruses: Detection, Characterization and Roles
Christopher I. Connell and Dominick P. Ralston (Editors)
2009. ISBN: 978-1-60692-965-0

Drug-Resistant Tuberculosis: Causes, Diagnosis and Treatments
Shui Nguy and Zhou K'ung (Editors)
2009. ISBN: 978-1-60876-055-8

Rabies: Symptoms, Treatment and Prevention
John G. Williamson (Editor)
2010. ISBN: 978-1-61668-250-7

Pseudorabies (Aujeszky's Disease) and Its Eradication
Tanja Mary and Laurent Claes (Editors)
2010. ISBN: 978-1-60741-655-5

Dengue Virus: Detection, Diagnosis and Control
Basak Ganim and Adam Reis (Editors)
2010. ISBN: 978-1-60876-398-6

Parvovirus B19 and Hematological Disorders in Children
Maysaa El Sayed Zaki
2010. ISBN: 978-1-61668-142-5
2010. ISBN: 978-1-61668-405-1 (E-book)

VIROLOGY RESEARCH PROGRESS

PARVOVIRUS B19 AND HEMATOLOGICAL DISORDERS IN CHILDREN

MAYSAA EL SAYED ZAKI

Nova Science Publishers, Inc.
New York

For permission to use material from this book please contact us:
Telephone 631-231-7269; Fax 631-231-8175
Web Site: http://www.novapublishers.com

Library of Congress Cataloging-in-Publication Data

El Sayed Zaki, Maysaa.
Parvovirus B19 and hematological disorders in children / Maysaa El Sayed Zaki.
p. ; cm.
Includes bibliographical references and index.
ISBN 978-1-62100-871-2 (softcover)
1. Parvovirus infections. 2. Hemolytic anemia--Complications. 3. Anemia in children--Complications. I. Title.
[DNLM: 1. Parvoviridae Infections--complications. 2. Parvovirus B19, Human--pathogenicity. 3. Child. 4. Hematologic Diseases--complications. WC 500 E49p 2010]
QR201.P33E4 2010
618.92'152--dc22

2010001752

Published by Nova Science Publishers, Inc. ✝ New York

CONTENTS

PREFACE

Parvovirus B19 is a childhood infectious disease of global proportions. It is the leading cause of erythema infectiousum, affecting mainly young age children. This book outlines the recent advances in understanding Parvovirus infection in children particularly associated hematological consequences. Also, it describes the associated complications by in utero infection such as hydropes fetalis and acute leukemia. It is emerged as occupational viral infection both to health care workers and school workers. It carries the danger of serious affection of recipients of blood and blood products transfusion. This book provides some glimpses into the current situation of this virus.

Parvovirus B19 PB19 is a very small (22 nm) non enveloped virus with a negative single-stranded DNA genome. It is the only member of the family parvoviridae known to be pathogenic in humans. The only natural host cell of PB19 is the human erythroid progenitors. In healthy individuals, the major presentation of PB19 infection is erythema infectiosum. However, in patients with underlying hemolytic disorders, infection is the primary cause of transient aplastic crisis. In immunosuppressed patients, persistent infection may develop that presents as pure red cell aplasia and chronic anemia. Inutero infection may result in hydrops fetalis or congenital anemia. Moreover, PB19 infection has been implicated in the development of childhood leukemia.

The virus can be transmitted by respiratory secretion, vertical transmission from mother to fetus and even by blood and blood products transfusion.. Some acute or chronic anemia may occur following the lysis of its target cell, the erythroid progenitors. Other clinical manifestations can be observed especially in immunocompromized patients.

Major advances in diagnosis of PB19 infection have taken place, including standardization of serological and DNA-based detection methodologies. As there is no reliable immunological method for antigen detection, hybridization or polymerase chain reaction (PCR) are needed for detecting viremia. Combined use of PCR and ELISA are optimal for diagnosis of PB19 infection.

Chapter 1

PARVOVIRUS B19 (PB19)

ABSTRACT

Parvovirus B19 (PB19) virion is DNA virus. It has a simple structure composed of only 2 proteins and a linear, single stranded DNA molecule. The non enveloped viral particles are 22 to 24 nm in diameter with icosahedral symmetry, and both empty and full capsids are visible by negative staining and electron microscopy. PB19 is an autonomous virus; its replication is dependant on cellular factors expressed transiently in the cell during the late S or early G2 phases of mitosis. The pathogenesis of PB19 is associated with erythroid tropism. The life cycle of PB19 like those of other non-enveloped DNA viruses. PB19 DNA has been found in the respiratory secretions at the time of viremia; transmission of infection via the respiratory route seems to be the most common route of spread of infection. Vertical transmission (from mother to fetus) occurs in one-third of cases involving serologically confirmed primary maternal infections. Furthermore, PB19 infection can be transmitted via blood derived products administrated parentally as the virus presents in high titers in the serum and is resistant to conventional heat treatments. Viremia is accompanied by constitutional symptoms of fever and malaise, and by erythroid progenitor cell depletion in the BM

Keywords: PB19 virology, epidemiology, pathogenesis.

INTRODUCTION

PB19 was first reported in 1974 while evaluating tests for hepatitis B virus surface antigen in sera of asymptomatic patients. The virus was designated as B19 after the code number of the serum sample (19) in panel B which contained the unexpected virus [1]. The human parvovirus B19 is now divided into three genotypes: type 1 (prototype), type 2 (A6- and LaLi-like), and type 3 (V9-like).

PB19 is a very small (22 nm) non enveloped virus with a negative single-stranded DNA genome without virion polymerase. The capsid has icosahedral symmetry (figure 1) and there is one serotype only. Since replication occurs only in erythrocyte precursors, PB19 is now classified as a member of the erythrocyte genus. It is the only one of parvovirus family which prove to infect man; it causes erythema infectiosum, aplastic anaemia and fetal infections including hydrops fetalis [2].

The development of PB19 specific serologic tests had led to the first report of symptomatic infection in humans. A report had described two soldiers who had a brief febrile illness with unknown origin. Subsequently, the simultaneous reports of PB19 as the etiologic agent for transient aplastic crisis (TAC) among patients with sickle cell disease and for erythema infectiosum among school children have established the association between parvovirus infection and these two disorders [4].

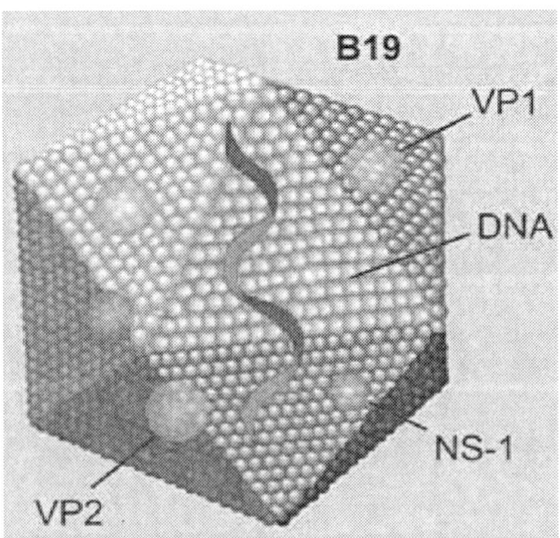

Figure 1. Parvoviruses are symmetrical icosahedral particles. They form small capsids and contain a DNA genome. The viral genome encodes only three proteins with known function, the nonstructural protein NS-1 and two capsid proteins VP1 and VP2.

PARVOVIRUS FAMILY (PARVOVIRIDAE)

Parvoviruses have been isolated from a wide range of hosts, from arthropods to humans. The family parvoviridae is divided into two subfamilies: Parvovirinae & Densivirinae. Densivirinae subfamily infects only invertebrates and is subdivided into 3 genera: Densovirus, Iteravirus and Contravirus. On the other hand, Parvovirinae infects both arthropods & humans. Parvovirinae similarly is subdivided into3 genera: Parvovirus, Dependovirus and Erythrovirus (table 1), [5].

Table 1. Taxonomic organization of Parvoviridae

Parvovirus	Minute virus of mice (MVM)	Mice	Sub clinical
	Feline parvovirus (FPV)	Cat	Enteritis, leucopenia, cerebellar ataxia
	Canine parvovirus (CPV)	Dog	Enteritis, myocarditis
	Porcine parvovirus (PPV)	Pigs	Reproductive failure
	Aleutian disease virus (ADV)	Mink	Pneumonitis
	Mink entritis virus (MEV)	Mink	Enteritis
Dependovirus	Adeno-associated viruses (AAV)	Humans & others	Unknown
Erythrovirus	B19	Man	Respiratory tract illness, aplastic crisis, erythema infectiosum/5thdisease & hydrops fetalis
	Simian parvovirus	Monkey	Anemia

PARVOVIRUS GENETIC STRUCTURE

The PB19 virion has a simple structure composed of only two proteins and a linear, single-stranded DNA molecule. The non-enveloped viral particles are 22 to 24 nm in diameter and show icosahedral symmetry, and often both empty and full capsids are visible by negative staining and Electron Microscope (EM) [6]. Mature infectious viral particles have a molecular weight of 5.6×10^6 Dalton [7]. The virion is composed of 60 copies of capsomers, and both negative and positive strands of DNA are packaged. X-ray crystallography has shown that the surface of PB19 is significantly different from those of other parvoviruses by lacking prominent spikes on the threefold icosahedral axes involved in host recognition and antigenicity. The limited DNA content and the absence of a lipid envelope make PB19 extremely resistant to physical inactivation. The virus is stable at 56°C for 60 minutes, and lipid solvents have no effect [7]. Inactivation of virus may be achieved by formalin, ß-propiolactone, and gamma irradiation [8].

PB19 virus is one of the stable viruses. The genetic diversity among PB19 virus isolates has been reported to be very low, with less than 1 to 2% nucleotide divergence in the whole genome, although full-length sequences are available only for a limited number of isolates [9-12]. Partial sequence data from different coding regions of the viral genome have confirmed this high degree of similarity with a larger number of isolates [13-17]. For example, sequence variation of the VP1/VP2 gene has been reported to be very low among PB19 virus isolates obtained from a single community-wide outbreak (0 to 0.6% base substitutions) and only slightly greater among PB19 virus isolates obtained from distinct epidemiological settings and geographical area, ranging between 0.5 and 4.8% for the most distant isolates [13]. PB19 virus genomes recovered from synovial tissue during persistent infection have also been reported to be very similar to those recovered from the same tissue during acute infection and to those recovered from blood or bone marrow [16]. However, some isolates obtained from patients with persistent B19 virus infection have been reported to exhibit a higher degree of variability in some parts of the genome, with the VP1 unique region being the most variable at both the DNA and protein levels with up to 4 and 8% divergence, respectively [15]. Although different genome types have been described based on restriction analysis of the PB19 virus genome [19, 20] sequence analysis has not allowed the identification of phylogenetic clusters with well-resolved nodes within the B19 viruses [11,17] .

Replication of a parvovirus entails double-stranded intermediate forms, which can be detected in tissue culture and clinical specimens by simple methods of DNA hybridization [11]. The transcription map of PB19 and the other erythroviruses differs markedly from that of other Parvoviridae, particularly in the use of a single promoter.

The viral genome encodes only three proteins of known function. The nonstructural protein (NS1) plays an important role in replication by providing endonuclease and helicase functions and initiating the complex transcription process. Moreover, it is cytotoxic to host cells inhibiting cellular division [5].

The two structural proteins, viral protein 1 (VP1) and viral protein 2 (VP2), arise from alternative splicing, so that VP1 is the same as VP2 except for an additional 226 amino acids at its amino terminal[25]. NS1 is encoded by genes on the left side of the genome and the two capsid proteins (VP1 and VP2) by genes on the right side[21]. The viral capsid formed of 60 capsomeres contains mainly VP2; while VP1 accounts for only about 5 percent of the capsid protein. Folding of the proteins creates α-helical loops that appear on the surface of the assembled capsids, where the host's immune system can recognize them as antigenic determinants. The region unique to VP1 is external to the capsid itself and contains many linear epitopes recognized by neutralizing antibodies [21].

In the last decade, new variants in the genus Erythrovirus have been described. Characterization of these isolates has shown a divergence of 10% or more between thevariants. Based on their sequences divergence, the genus has been divided into three genotypes: 1 (B19), 2 (A6 and LaLi), and 3 (V9) [22]. Few clinical and molecular data on genotype-3 strains were available until West Africa was identified as an endemic region for genotype-3 infection [23]. Parsyan et al. 2007 [24], in a study with genotype-3 samples from Ghana, observed two different clusters and proposed to recognize two subtypes of genotype 3, called 3a (B19/3a) and 3b (B19/3b).

PB19 is an autonomous virus; not requiring the presence of a helper virus, its replication is dependant on cellular factors expressed transiently in the cell during the late S or early G2 phases of mitosis. The necessary factors may only be produced at specific stages of cellular differentiation, thus PB19 requires dividing cells for replication, but not all dividing cells are susceptible[5].

Parvovirus B19 is dependent on mitotically active erythroid progenitor cells for replication. In human clonal progenitor studies there is marked inhibition of erythroid colony formation by CFU-E and BFU-E, with virtually no e€ect on colony formation from CFU-GM [25]. Susceptibility to parvovirus B19 increases with differentiation, and the pluripotent stem cell appears to be spared. In erythroid progenitors the virus is directly cytotoxic, with infected cells showing the ultrastructural [26] and biochemical [27, 28] features typical of apoptosis [26]. In addition, it has been shown in vitro that non-structural protein expression induces expression of the inflammatory cytokine, interleukin 6, which may contribute to disease pathology[27].

PB19 infected bone marrow cultures are characterized by the presence of giant pronormoblasts or `lantern cells': early erythroid cells with cytoplasmic vacuolization, immature chromatin and large eosinophilic nuclear inclusion bodies. The light microscopic findings are also seen in the bone marrow of infected patients, with occasional giant pronormoblasts even seen in the peripheral circulation.

In vitro, autonomous PB19V replication is limited to human erythroid progenitor cells and in a small number of erythropoietin-dependent human megakaryoblastoid and erythroid leukemic cell lines. It was reported report that the failure of B19V DNA replication in nonpermissive [29] cells can be overcome by adenovirus infection. More specifically, the replication of B19V DNA in the [29] cells and the production of infectious progeny virus were made possible by the presence of the adenovirus E2a, E4orf6, and VA RNA genes that emerged during the transfection of the pHelper plasmid. Using this replication system, the terminal resolution site was identified and the nonstructural protein 1 (NS1) binding site on the right terminal palindrome of the viral genome, which is composed of a minimal origin of replication spanning 67 nucleotides. Plasmids or DNA fragments containing an NS1 expression cassette and this minimal origin were able to replicate in both pHelper-transfected [29] cells and B19V-semipermissive UT7/Epo-S1 cells. These results have important implications for understanding of native B19V infection [30].

The pathogenesis of PB19 is associated with erythroid tropism for P blood group. The erythroid tropism of PB19 is due to the tissue-specific expression of the cellular receptor for parvovirus PB19 [31]. Virus host cell binding is mediated through globoside, a neutral glycosphingolipid found predominantly on erythroid cells or their progenitors, where it is known as the blood group P antigen. Only red cells containing blood group P antigen can be haemagglutinated, and bone marrow from patients that do not have P on their erythrocytes cannot be infected with parvovirus B19 in vitro [32].

The P blood group was discovered in 1927 by Landsteiner and Levine as part of studies to identify new human blood group antigens by immunizing rabbits with human erythrocytes. It is now known that the P blood group system contains two common antigens, P1 and P, and the rarer antigen, Pk. Red cells of individuals with blood group P1 phenotype have P1 and P antigens; individuals with P1 k phenotype have P1 and Pk antigens; individuals with P2 phenotype have P antigen alone; and individuals with the rare p phenotype (previously known as Tjaÿ) lack all three antigens. The P1 and P2 phenotypes are both very common and account for virtually 100% of individuals in all ethnic groups studied. The Pk and p phenotypes are very rare,with estimates of the prevalence of the p phenotype in the general population being 1:200 000, but more frequent in Japan, Sweden, and among the Amish in the USA. In seroprevalence studies among the Amish in Ohio, USA, it was shown that although B19 is prevalent in the Parvovirus B19 infection [33]community, individuals with p phenotype had no evidence of previous infection with parvovirus B19 [34], confirming the importance of globoside in B19 infection.

P antigen is found on erythroid progenitors, erythroblasts, and megakaryoblasts and megakaryocytes. It is also present on endothelial cells, which may be targets of viral infection involved in the pathogenesis of transplacental transmission, possibly vasculitis and the rash of fifth disease, and on fetal myocardial cells. Later on, it was found that mature human red blood cells (RBCs), which express high levels of P antigen, but not α5β1 integrins, bind parvovirus B19 but do not allow viral entry. In contrast, primary human erythroid progenitor cells express high levels of both P antigen and α5β1 integrins and allow β1 integrin-mediated entry of parvovirus B19. Thus, in a natural course of infection, RBCs are likely exploited for a highly efficient systemic dissemination of parvovirus B19 [35].

The life cycle of PB19, like those of other non-enveloped DNA viruses, includes binding of the virus to host cell receptors, internalization, translocation of the genome to the host nucleus, DNA replication, RNA transcription, assembly of capsids and packaging of the genome, and finally cell lyses with release of the mature virions [2].

Epidemiology of PB19

PB19 is a global and common infectious pathogen in humans. The prevalence of IgG antibodies directed against PB19 ranges from 2 to 15% in children 1 to 5 years old, 15 to 60% in children 6 to 19 years old, 30 to 60% in adults, and more than 85% in the geriatric population [36]. Women of childbearing age show an annual seroconversion rate of 1.5%[2]. Although antibody is prevalent in the general population, viremia or presence of viral DNA is rare. PB19 virus infection is endemic throughout the year in temperate climates, but there is a seasonal increase in frequency in late winter, spring & early summer months. Rates of infection may rise to an epidemic level every 4-5 years manifested by outbreaks of erythema infectiosum or PB19 induced aplastic crisis [5].

Transmission of PB19

PB19 DNA has been found in the respiratory secretions at the time of viremia; transmission of infection via the respiratory route seems to be the most common route of spread of infection.

Vertical transmission (from mother to fetus) occurs in one-third of cases involving serologically confirmed primary maternal infections[37-41]. Foetal capillary endothelium in placental villi can be an additional target of productive B19 virus infection. Infection of placental endothelial cells may lead to a structural and functional damage critical both for altering maternal-foetal blood exchanges and for spreading the infection to the foetus, possibly concurring to the development of foetal hydrops and intrauterine foetal death [39]. Nosocomial transmission has been described infrequently and transmission has also been reported among staff in laboratories handling native virus [40].

Furthermore, PB19 infection can be transmitted via blood derived products administrated parentrally as the virus presents in high titers in the serum and is resistant to conventional heat treatments. The risk of infection using single-donor blood products is reportedly varied but is probably low [38].

Several reports have proven that PB19 can be transmitted through blood transfusions and plasma-derived products[41,42]. Screening of blood donations for the presence of B19 DNA is not routine[43], despite the fact that this virus is highly resilient and, like the hepatitis A virus, can withstand denaturation, even at high temperatures[44].

Normally, in the general population, a continuous increase in PB19 seroreactivity is observed with age, with a seroprevalence for adults over 60 that reaches about 72 %. Interestingly, the haemophilic population that was treated pre-1984 with non-inactivated clotting factor concentrates had an increased seroprevalence of 98 % [45, 46].

Pathogenesis of PB19 Infection

Human PB19 infects predominantly erythroid precursor cells, leading to inhibition of erythropoiesis, thus it is called erythrogenic virus. This erythroid cell damage is mediated by the viral NS1 through an apoptotic mechanism[47]..

The deficiencies of appropriate immune responses to B19 impair viral elimination in vivo, which results in enlargement of B19-infected erythroid-lineage cells. The B19-associated damage of erythroid lineage cells is due to cytotoxicity mediated by viral proteins. B19-infected erythroid-lineage cells show apoptotic features, which are thought to be induced by the non-structural protein, NS1, of B19. In addition, B19 infection induces cell cycle arrests at the G (1) and G(2) phases. The G(1) arrest is induced by NS1 expression prior to apoptosis induction in B19-infected cells, while the G(2) arrest is induced not only by infectious B19 but also by UV-inactivated B19, which lacks the ability to express NS1[48].

The presence of a specific cellular receptor is thought to be necessary for susceptibility to viral infection. The erythrocyte P antigen is suggested to be the cellular receptor for PB19. PB19 &PB19 VP2 both bind directly to P antigen, and in tissue culture either excess P antigen or anti-P monoclonal antibody can protect erythroid progenitors from infection with PB19, thus demonstrating that P antigen is the PB19 receptor. In addition, individuals who are genetically lacking P antigen are naturally resistant to PB19 infection and can't be infected even in the presence of high concentrations of the virus (38, 49). In spite of presence of P antigen on megakaryocytes, endothelial cells, and fetal myocytes; none of these cell types have been shown to be permissive for PB19 replication [2].

Recent studies show that low oxygen pressure in the BM and fetal tissues leads to higher yields of infectious PB19 progeny and to a higher level of viral transcription than observed under normoxia [50, 51].

Viremia occurs during the first week of infection, accompanied by constitutional symptoms of fever and malaise, and by erythroid progenitor cell depletion in the BM. At the height of the viremia, a precipitous drop in the reticulocytic count occurs followed by anemia, which is rarely clinically apparent in healthy patients but can cause serious anemia if the red blood cell count is already low. The reduction in the reticulocytic count is occasionally accompanied by leucopenia and thrombocytopenia[4].

Immune Response to PB19

The unique region of the capsid protein VP1 (VP1u) of human PB19 elicits a dominant immune response and has a phospholipase A (2) activity, which is necessary for the infection. When using a monoclonal antibody against the N-terminal portion of VP1u, authors revealed that this region rich in neutralizing epitopes that is not accessible in native capsids. The anti-VP1u antibody, although unable to bind free virus or to block virus attachment to the cell, has neutralizing action for virus. This finding is revealing that the exposure of the epitope and the subsequent virus neutralization occur only after receptor attachment [52]. The majority of patients who develop immunoglobulin (Ig) G antibodies against the capsid proteins VP1 and VP2 have life-long protection against reinfection. Most virus neutralizing antibodies that offer protection are directed against VP1u. Immunoglobulin M antibodies are mainly directed against VP2-specific epitopes. These antibodies can be present for only a rather short period of 2 to 10 weeks after acute infection [53].

The appearance of PB19 specific IgM antibodies in the serum in the second week after inoculation corresponds with clearance of the viremia. In the third week after inoculation, specific IgG antibodies appear in the serum, and the rash of erythema infectiosum and arthropathy develop. The appearance of the rash corresponds with the development of IgG antibodies and occurs after the viremia has cleared. That's to say that the rash of erythema infectiosum signifies that the virus can no longer be transmitted [4]

Figure (2) describes both T- and B- lymphocytes cell response to parvovirus B19 infection. Upon PB19 infection, B cells divide to produce plasma cells and memory cells. Plasma cells secrete IgM antibodies that are specific for both conformational (N) and linear (D) epitopes of PB19, which are usually detectable around 7 days post-infection. PB19-specific IgG is detectable about 15 days post-infection and is directed initially against both linear and conformational epitopes of the capsid proteins (VP1 and VP2) and, to a lesser extent, against NS1, but declines against linear epitopes of the proteins in a time-dependent manner. Infection by PB19 most likely confers lifelong protection on the host, due to the development of memory B cells that are specific for the conserved regions of the PB19 capsid and also linear regions of the VP1 protein [54]. Antigen-presenting cells (APC) process the virus and displayP B19 peptides on their surface combined with MHC I to Th cells. These Th cells then secrete cytokines that play a role in mediating anti-virus immunity and may also be associated with pathogenesis of B19 infection (e.g. IL6 is associated with RA).

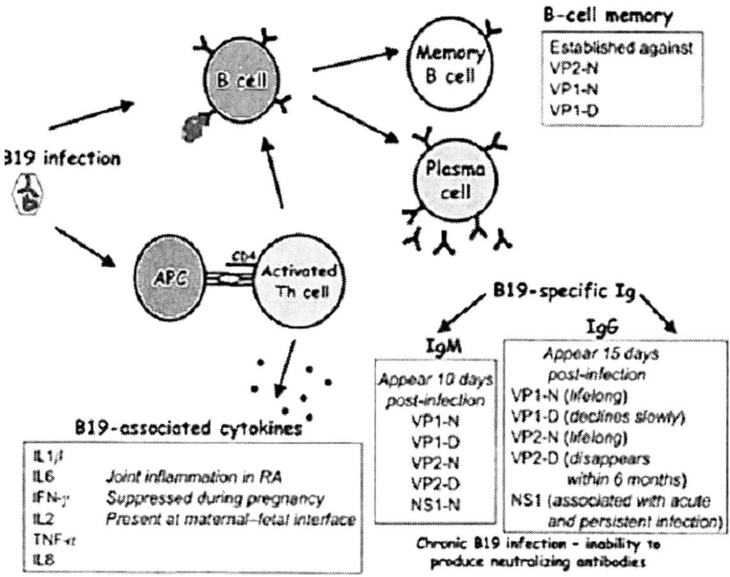

Figure 2. Schematic depiction of T- and B-cell response to parvovirus B19 infection.

Cell-mediated immunity to B19 has not been studied extensively; this is due primarily to the fact that the humoral response was thought to be most important in combatting B19 infection. Indeed, initial attempts to demonstrate specific T-cell proliferative responses to B19 were unsuccessful [55] and, for some time, this work supported the prevailing theory that neutralizing antibody production was the major mechanism of immunity in B19. In 1996, *ex vivo* B19-specific CD4$^+$ T-cell responses were first detected against *E. coli*-expressed VP1, VP2 and NS1 antigens [56]. T-cell responses of 16 individuals were analysed (ten seropositive and six seronegative blood donors), none of whom had any evidence of acute infection. The majority (90 %) of seropositive donors who were stimulated *ex vivo* by VP2 displayed specific T-cell responses, with 80 % displaying VP1-specific responses. There was no significant difference in T-cell proliferation for NS1 between seropositive and seronegative individuals. Upon inclusion of mAbs that were specific for class I and class II HLA, it was found that HLA class II-specific antibodies inhibited T-cell proliferation, thus indicating that the effector T-cell population of B19 are CD4$^+$ cells. Subsequent peripheral blood mononuclear cell (PBMC) depletion of either CD4$^+$ or CD8$^+$ T cells and stimulation of the remaining population confirmed this observation.

More recently, significant *ex vivo* T-cell reactivity was observed in PBMCs of recently and remotely infected individuals by using a B19 candidate vaccine [57] and also the B19 recombinant proteins, VP1 and VP2[62]. T cells from recently infected individuals responded strongly to the B19 capsids, giving a mean T-cell stimulation index (SI) of 36 [58]. Blood donors with past infections gave comparable rates of T-cell stimulation. Seronegative individuals had SI values of about 3.3 and this study also showed that the responding population of T cells were $CD4^+$. Although There is no difference in T-cell responses to NS1 in seronegative and seropositive individuals, significant responses to this antigen have been reported in recently infected individuals and patients who developed chronic arthropathy following B19 infection [60, 61]. T-cell responses to NS1 were not seen in healthy individuals with past B19 infection, except for two individuals who were also NS1 IgG-seropositive.

Cellular immune response to an epitope of NS1 that is specifically recognized by $CD8^+$ T cells was investigated by using major histocompatibility complex tetrameric complex binding [61]. The response of 21 individuals to this epitope was examined in healthy volunteers and human immunodeficiency virus (HIV)-1 infected adults and children. Sixteen of the volunteers were HLA-matched (HLA B35) and six were mismatched. Sixty-three per cent of matched individuals displayed specific $CD8^+$ T-cell responses. Seventy-two per cent of matched individuals in the same cohort exhibited specific T-cell responses by using an interferon γ(IFN-γ) ELISpot assay [59]. The level of B19-specific $CD8^+$ T cells was similar among healthy and HIV-infected individuals. The results presented in this report showed the important cellular role of cytotoxic T cells in combating PB19 infection [60]. PB19-specific T-cell responses may now represent a novel method for confirming past B19 infection [62].

Recent evidence shows the importance of evaluating T-cell responses in understanding the nature of B19 infection. There has been identified an AIDS patient with persistent B19 infection who showed an initial remission of B19 infection [63]. This remission was evident despite the lack of a specific antibody response, thus indicating a role for cellular immunity in combatting B19 infection. NS1-reactive lymphocytes have been detected in two B19-seronegative individuals who were exposed to the virus, indicating a possible subclinical B19 infection or perhaps a loss of antibodies against capsid proteins [62]. The importance of cellular immunity in B19 was further emphasized. The presence of PB19-specific $CD8^+$ T-cell responses identified two healthy adults and two HIV-1-infected patients who were seronegative for

B19, with specific T-cell responses against B19 by either IFN-γ ELISpot or tetramer binding studies, thus implying the presence of a cellular response in the absence of a humoral response [63].

Significant T-cell transcriptional activation has been reported in a patient with acute B19 infection, causing increased levels of interleukin (IL) 1ß, IL6 and IFN-γ mRNA [64]. A subsequent study that analysed the sera of patients who were infected acutely by B19 showed that although IL1ß, IL6, IFN-γ and tumour necrosis factor α(TNF-α) were secreted during the acute phase of infection, increased levels of both IFN-γ and TNF-α persisted and were detectable 2–37 months later, during a follow-up study. It has also been suggested that cytokine genetic polymorphisms may, in some way, affect the development of symptoms during B19 infection. To date, the transforming growth factor ß (TGF-ß) allele has been associated with skin rash at acute infection and the IFN-γ allele has been associated with NS1 antibody development [65]. In a study of recently infected children, it was shown that although strong T-cell proliferative responses were evident to both capsid proteins, production of the T helper (Th1) cytokine IFN-γ, but not of IL2, was impaired when compared to convalescent adults [66]. In addition, *ex vivo* production of IFN-γ and IL2 that was observed in B19-seropositive pregnant women was lower than observed previously for healthy, non-pregnant individuals, suggesting a possible dimunition of the maternal anti-virus immune response that may subsequently increase the risk of fetal B19 infection [67]. Expression of the non-structural protein NS1 causes the production of increased levels of the inflammatory cytokine IL6 in a number of cell lines, including haematopoietic cell lines and human umbilical vein endothelial cells[68]. IL6 is known to be involved in synovial cell proliferation and, in addition, high levels of IL6, along with other inflammatory cytokines, have been found in inflamed joints of patients with RA, which would suggest an association between IL6 production and the joint manifestations that are observed with PB19 infection [69]. IL6 involvement in RA is supported by the fact that antibodies against IL6 cause inhibition of RA manifestations [68]. As well as increased IL6 production, high levels of IFN-γ, TNF-α and IL8 have been detected in the sera of infants with B19-associated acute myocarditis [69]. IL2 production at the maternal–fetal interface in women who seroconverted to B19 during pregnancy is thought to determine the outcome of the pregnancy, with high levels of IL2 on the fetal side being associated with pregnancies that result in a poor outcome .

Virologic, immunologic and clinical course following PB19 infection is shown in figure (3), [70].

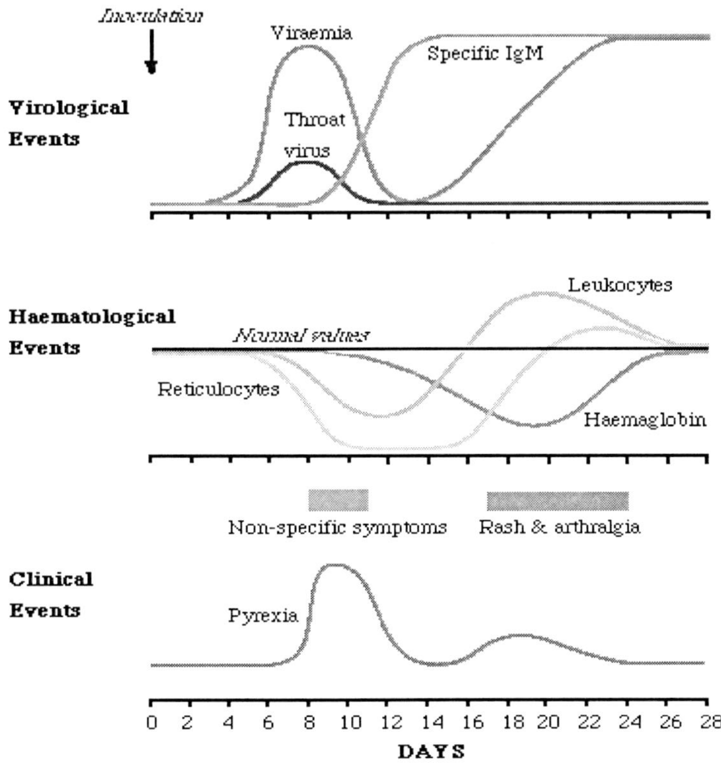

Figure (3). Virologic, immunologic and clinical course following PB19 infection (69).

Conclusion

Parvovirus B19 (PB19) virion has a simple structure composed of only 2 proteins and a linear, single stranded DNA molecule. The non enveloped viral particles are 22 to 24 nm in diameter and show icosahedral symmetry, and often both empty and full capsids are visible by negative staining and electron microscopy.1

Children are probably at highest risk of acquiring PB19 infection because they may not yet have encountered the infection and thus lack immunity. Parvovirus exhibits a marked tropism to human bone marrow, replicating exclusively in erythroid progenitor cells. In immunocompetent children, PB19 is the cause of er ythema infectiosum, whereas arthropathy is infrequently observed. In immunocompromised hosts such as cancer patients, PB19 infection may persist and lead to chronic anemia, red cell aplasia, and, less frequently, thrombocytopenia, neutropenia, and pancytopenia. Healthy hosts are able to clear the virus within weeksafter infection

References

[1] Cossart YE, Field AM, Cant B, et al. Parvovirus-like particles in human sera. *Lancet.* 1975, i: 72-3.

[2] Heegaard ED and Brown KE Human Parvovirus B19. *Clin. Microb. Rev.* 2002, 15:485-505.

[3] Broliden K, Tolfvenstam T & Norbeck O. Clinical aspects of parvovirus B19 infection. *J. Intern. Med.* 2006; 260: 285–304

[4] Sabella C and Goldfarb J. Parvovirus B19 Infections. *Am. Fam. Physician.* 1999, 60:1455-60.

[5] Cubie HA. Viral pathogens an associated diseases (Parvoviruses) in Medical Microbiology.16th edition. Greenwood - Churchill Livingstone. 2002, 448-54.

[6] Berns KI. Parvoviridae: the viruses and their replication, *In* BN Fields, DM Knipe, PM Howley, et al. Fields virology. 1996, Lippincott-Raven, Philadelphia, Pa. 2173-97.

[7] Kerr JR. The Parvoviridae; an emerging virus family. *Infect. Dis. Rev.* 2000, 2: 99-109.

[8] Cohen B J and Brown K E. Letter, *J. Infect.* 1992, 24:113-114.

[9] Hicks, K. E., R. C. Cubel, B. J. Cohen, and J. P. Clewley. 1996. Sequence analysis of a parvovirus B19 isolate and baculovirus expression of the non-structural protein. *Arch. Virol.* 141:1319-1327.

[10] Hokynar, K., J. Brunstein, M. Soderlund-Venermo, O. Kiviluoto, E. K. Partio, Y. Konttinen, and K. Hedman. 2000. Integrity and full coding sequence of B19 virus DNA persisting in human synovial tissue. *J. Gen. Virol.* 81:1017-1025

[11] Lukashov, V. V., and J. Goudsmit. 2001. Evolutionary relationships among parvoviruses: virus-host coevolution among autonomous primate parvoviruses and links between adeno-associated and avian parvoviruses. *J. Virol.* 75:2729-2740.[

[12] Shade, R. O., M. C. Blundell, S. F. Cotmore, P. Tattersall, and C. R. Astell. 1986. Nucleotide sequence and genome organization of human parvovirus B19 isolated from the serum of a child during aplastic crisis. *J. Virol.* 58:921-936.

[13] Erdman, D. D., E. L. Durigon, Q. Y. Wang, and L. J. Anderson. 1996. Genetic diversity of human parvovirus B19: sequence analysis of the VP1/VP2 gene from multiple isolates. *J. Gen. Virol.* 77:2767-2774

[14] Gallinella, G., S. Venturoli, G. Gentilomi, M. Musiani, and M. Zerbini. 1995. Extent of sequence variability in a genomic region coding for capsid proteins of B19 parvovirus. *Arch. Virol.* 140:1119-1125

[15] Hemauer, A., A. von Poblotzki, A. Gigler, P. Cassinotti, G. Siegl, H. Wolf, and S. Modrow. 1996. Sequence variability among different parvovirus B19 isolates. *J. Gen. Virol.* 77:1781-1785

[16] Umene, K., and T. Nunoue. 1993. Partial nucleotide sequencing and characterization of human parvovirus B19 genome DNAs from damaged human fetuses and from patients with leukemia. *J. Med. Virol.*

[17] Mori, J., P. Beattie, D. W. Melton, B. Cohen, and J. P. Clewley. 1987. Structure and mapping of the DNA of human parvovirus B19. *J. Gen. Virol.* 68:2797-2806.

[18] Morinet, F., J. D. Tratschin, and Y. Perol. 1986. Comparison of 17 isolates of the human parvovirus B19 by restriction enzyme analysis. *Arch. Virol.* 90:165-172

[19] Umene, K., and T. Nunoue. 1990. The genome type of human parvovirus B19 strains isolated in Japan during 1981 differs from types detected in 1986 to 1987: a correlation between genome type and prevalence. *J. Gen. Virol.* 71:983-986.

[20] Umene, K., and T. Nunoue. 1991. Genetic diversity of human parvovirus B19 determined using a set of restriction endonucleases recognizing four or five base pairs and partial nucleotide sequencing: use of sequence variability in virus classification. *J. Gen. Virol.* 72:1997-2001. [20]. Lukashov, V. V., and J. Goudsmit. 2001. Evolutionary relationships among parvoviruses: virus-host coevolution among autonomous primate parvoviruses and links between adeno-associated and avian parvoviruses. *J. Virol.* 75:2729-2740.[

[21] Young NS and Brown KE (2004): Parvovirus B19. *The New England Journal of Medicine.*350:586-97.

[22] Servant A, Laperche S, Lallemand F, Marinho V, De Saint Maur G, Meritet JF, Garbarg-Chenon A (2002) Genetic diversity within human erythrovirus: identification of three genotypes. *J. Virol.* 76:9124–9134

[23] Candotti D, Etiz N, Parsyan A, Allain JP (2004) Identification and characterization of persistent human erythrovirus infection in blood donor samples. *J. Virol.* 78:12168–12178

[24] Parsyan A, Szmaragd C, Allain JP, Candotti D (2007) Identification and genetic diversity of two human parvovirus B19 genotype 3 subtypes. *J. Gen. Virol.* 88:428–431

[25] Mortimer PP, Humphries RK, Moore JG et al. A human parvovirus-like virus inhibits haematopoietic colony formation in vitro. *Nature.* 1983; 302: 426±429.

[26] Morey AL, Ferguson DJ & Fleming KA. Ultrastructural features of fetal erythroid precursors infected with parvovirus B19 in vitro: evidence of cell death by apoptosis. *Journal of Pathology.* 1993; 169: 213±220.

[27] Moffatt S, Yaegashi N, Tada K et al. Human parvovirus B19 nonstructural (NS1) protein induces apoptosis in erythroid lineage cells. *J. of Virol.* 1998; 72: 3018-3028.

[28] Sol N, Le Junter J, Vassias I et al. Possible interactions between the NS-1 protein and tumour necrosisfactor alpha pathways in erythroid cell apoptosis induced by human parvovirus B19. *Journal of Virology.* 1999; 73: 8762±8770.

[29] Moffatt S, Tanaka N, Tada K et al. A cytotoxic nonstructural protein, NS1, of human parvovirus B19induces activation of interleukin-6 gene expression. *Journal of Virology.* 1996; 70: 8485±8491.

[30] Guan W, Wong S, Zhi N, Qiu J .The genome of human parvovirus b19 can replicate in nonpermissive cells with the help of adenovirus genes and produces infectious virus. *J. of Virol,* 2009, 83, 18: 9541-9553.

[31] Brown KE, Anderson SM & Young NS. Erythrocyte P antigen: cellular receptor for B19 parvovirus. *Science.* 1993; 262: 114±117.

[32] Brown KE, Hibbs JR, Gallinella G et al. Resistance to parvovirus B19 infection due to lack of virus receptor (erythrocyte P antigen). *New England Journal of Medicine.* 1994; 330: 1192±1196.

[33] Rouger P, Gane P & Salmon C. Tissue distribution of H, Lewis and P antigens as shown by a panel of 18 monoclonal antibodies. Revue Françoise de Transfusion et Immuno-heÂmatologie 1987; 30: 699±708.

[34] Weigel-kelley K A.; Mervin C Y; Arun S. α5β1 integrin as a cellular coreceptor for human parvovirus B19: requirement of functional activation of β1 integrin for viral entry. *Blood.* 2003, 102: 12/ 3927-3933.

[35] Kelly HA, Siebert D, Hammond R, et al. The age-specific prevalence of human parvovirus immunity in Victoria, Australia compared with other parts of the world. *Epidemiol. Infect.* 2000, 124:449-457

[36] Public Health Laboratory Service Working Party on Fifth Disease. Prospective study of human parvovirus (B19) infection in pregnancy. *Br. Med. J.* 1990,300:1166-70.

[37] Mihneva Z. Clinical manifestations, risks and trends regarding human Parvovirus B19 infection. *Med. Rev.* 2005, 41:5-14.

[38] Pasquinelli G, Bonvicini F, Foroni L, Salfi N, Gallinella G. Placental endothelial cells can be productively infected by Parvovirus B19. *J. Clin. Virol.* 2009; 44:1/33-8

[39] Miyamoto K, Ogami M, Takahashi Y, et al. Outbreak of human parvovirus B19 in hospital workers. *J. Hosp. Infect.* 2000, 45:238-241

[40] Prowse, C., Ludlam, C. A. & Yap, P. L. Human parvovirus B19 and blood products. *Vox Sang.* 1997, 72, 1–10

[41] Santagostino E, Mannucci P M, Gringeri A, Azzi A, Morfini M, Musso R, Santoro R & Schiavoni M. Transmission of parvovirus B19 by coagulation factor concentrates exposed to 100 degrees C heat after lyophilization. *Transfusion.* 1997, 37, 517–522.

[42] Blümel, J., Schmidt, I., Willkommen, H. & Löwer, J. Inactivation of parvovirus B19 during pasteurization of human serum albumin. *Transfusion.* 2002, 42: 1011–1018.

[43] Santagostino E, Mannucci P M, Gringeri A, Azzi A & Morfini M. Eliminating parvovirus B19 from blood products. *Lancet.* 1994, 343, 798. 798.

[44] Williams M D, Cohen B J, Beddall A C, Pasi K J, Mortimer P P& Hill, F G H. Transmission of human parvovirus B19 by coagulation factor concentrates. *Vox Sang.*1990,58: 177–181.

[45] Eis-Hübinger A M, Oldenburg J, Brackmann H H, Matz B & Schneweis K E. The prevalence of antibody to parvovirus B19 in hemophiliacs and in the general population. *Zentbl Bakteriol.* 1996, 284: 232–240.

[46] Morita E, Nakashima A, Asao H, et al. Human Parvovirus B19 Non Structural Protein (NSI) Induces Cell Cycle Arrest at G1 Phase. J. *of Virology.* 2003, 77:2915-21.

[47] Chisaka H, Morita E, Yaegashi N, Sugamura K. Parvovirus B19 and the pathogenesis of anaemia. *Rev. Med. Virol.* 2003; 13(6):347-59.

[48] Kevin E, Brown, Jonathan R, et al. Resistance to Parvovirus B19 Infection Due to Lack of Virus Receptor (Erythrocyte P Antigen). *NEJM.* 1994, 330:1192-96.

[49] Caillet-Fauquet P, Draps ML, Di Giambattista M, et al. Hypoxia enables B19 erythrovirus to yield abundant infectious progeny in a pluripotent erythroid cell line. *J. Virol. Methods.* 2004, 121: 145-53.

[50] Pillet S, Le Guyader N, Hofer T, et al. Hypoxia enhances human B19 erythrovirus gene expression in primary erythroid cells. *Virology*. 2004, 327: 1-7.

[51] Ros C, Gerber M, Kempf C. Conformational changes in the VP1-unique region of native human parvovirus B19 lead to exposure of internal sequences that play a role in virus neutralization and infectivity. *J. Virol.* 2006; 80: 12017–12024.

[52] Modarows S, Dorsch S. Antibody responses in parvovirus B19 infected Nigro, G., Bastianon, V., Colloridi, V., Ventriglia, F., Gallo, P., D'Amati, G., Koch, W. C. & Adler, S. P. Human parvovirus B19 infection in infancy associated with acute and chronic lymphocytic myocarditis and high cytokine levels: report of 3 cases and review. *Clin. Infect. Dis.*2000, 31, 65–69.

[53] Kurtzman G J, Ozawa K, Cohen B J, Hanson G, Oseas R & Young N S. Chronic bone marrow failure due to persistent B19 parvovirus infection. *NEJM.* 1987, 317: 287–294

[54] Corcoran A, Doyle S Advances in the biology, diagnosis and host-pathogen interactions of parvovirus B19. *J. Med. Microbiol.* 2004 ;53: 6/459-75.

[55] von Poblotzki A, Gerdes C, Reischl U, Wolf H & Modrow S. Lymphoproliferative responses after infection with human parvovirus B19. *J Virol.* 1996, 70: 7327–7330

[56] Franssila R, Hokynar K. & Hedman K. T helper cell-mediated in vitro responses of recently and remotely infected subjects to a candidate recombinant vaccine for human parvovirus B19. *J Infect Dis.* 2001, 183: 805–809.

[57] Corcoran A, Doyle S, Waldron D, Nicholson A & Mahon B P. Impaired gamma interferon responses against parvovirus B19 by recently infected children. *J. Virol.* 2000, 74: 9903–9910

[58] Kerr J R, Barah F, Mattey D L, Laing I, Hopkins S J, Hutchinson I V & Tyrrell D A J. Circulating tumour necrosis factor-α and interferon-γ are detectable during acute and convalescent parvovirus B19 infection and are associated with prolonged and chronic fatigue. *J. Gen. Virol.* 2001, 82 : 3011–3019

[59] Kerr J R, Curran M D, Moore J E, Coyle P V & Ferguson W P Persistent parvovirus B19 infection. *Lancet.* 1995, 345: 1118-22

[60] Tolfvenstam T, Papadogiannakis N, Norbeck O, Petersson K & Broliden K. Frequency of human parvovirus B19 infection in intrauterine fetal death. *Lancet.* 2001, 357: 1494–1497

[61] Chen M Y, Hung C C, Fang C T & Hsieh S M. Reconstituted immunity against persistent parvovirus B19 infection in a patient with acquired immunodeficiency syndrome after highly active antiretroviral therapy. *Clin. Infect. Dis.* 2001, 32: 1361–1365

[62] Mitchell L A, Leong R & Rosenke K A. Lymphocyte recognition of human parvovirus B19 non-structural (NS1) protein: associations with occurrence of acute and chronic arthropathy? *J. Med. Microbiol.* 2001, 50: 627–635

[63] Wagner A D, Goronzy J J, Matteson E L & Weyland C M. Systemic monocyte and T-cell activation in a patient with human parvovirus B19 infection. *Mayo Clin. Proc.* 1995, 70: 261–265

[64] Kerr J R, McCoy M, Burke B, Mattey D L, Pravica V & Hutchinson I V. Cytokine gene polymorphisms associated with symptomatic parvovirus B19 infection. *J. Clin. Pathol.* 2003,56: 725–727.

[65] Corcoran A, Mahon B P & Doyle S. B cell memory is directed towards conformational epitopes of parvovirus B19 capsid proteins and the VP1-unique region. *J. Infect. Dis.* 2004, 189: 1873-80.

[66] Moffatt, S., Tanaka, N., Tada, K., Nose, M., Nakamura, M., Muraoka, O., Hirano, T. & Sugamura, K. (1996). A cytotoxic nonstructural protein, NS1, of human parvovirus B19 induces activation of interleukin-6 gene expression. *J. Virol.* 70, 8485–8491.

[67] Bataille R, Barlogie B, Lu Z Y, Rossi J F, Lavabre-Bertrand T, Beck T, Wijdenes J, Brochier J & Klein B. Biologic effects of anti-interleukin-6 murine monoclonal antibody in advanced multiple myeloma. *Blood.* 1995, 86: 685–691.

[68] Nigro G, Bastianon V, Colloridi V, Ventriglia F, Gallo P, D'Amati G, Koch W C & Adler S P. Human parvovirus B19 infection in infancy associated with acute and chronic lymphocytic myocarditis and high cytokine levels: report of 3 cases and review. *Clin. Infect. Dis.* 2000,31: 65–69.

[69] Brown KE and Young NS. Parvovirus B19 in human disease. *Ann. Rev. Med.* 1997, 48:59-67.

Chapter 2

CLINICAL ASPECTS OF PB19 INFECTION IN IMMUNOCOMPETENT PATIENTS

ABSTRACT

Parvovirus B19 infections are associated with different clinical manifestations that vary from asymptomatic to severe symptoms. The main clinical manifestations are erythema infectiosum, acute polyarthralgia syndrome in adults, hydrops fetalis, spontaneous abortion, and stillbirth. It is reported that the hematological consequences of B19 infection arise due to direct cytotoxic effect on progenitors in the bone marrow with interruption of erythroid production. PB19 infection has been associated with a wide variety of clinical manifestations and some clinical features of B19 infection such as anemia or rash can be common to other pathogens, a specific laboratory identification of B19 is required and any diagnostic tool must consider both the type of pathology and the type of patient. In immunocompetent individuals, virologic and serologic testing are complementary. Children are probably at highest risk of acquiring parvovirus B19 infection, as they may not have yet encountered the infection and thus lack immunity. Prevention of disease by isolating susceptible individuals is impractical since infections may be subclinical and symptomatic individuals are infectious before any sign of illness. Theoretically, susceptible individuals with chronic haemolytic anemia or immunocompromised children could be temporarily protected by the administration of human immunoglobulin.

Keywords: PB19, clinical symptoms, immunocompetents patients.

Introduction

The clinical and pathomorphological patterns of PB19-associated diseases are the result of a balance between virus, host target cells and immune response. It is a characteristic feature of PB19 that in patients with various other preexisting diseases, e.g., many hemolytic anemias, immune complex-mediated vasculitic disorders, and primary or secondary immunodeficiencies, the underlying diseases can be triggered, aggravated or complicated by severe organ manifestations. Identification of PB19 by means of routine histology and immunohistology is only given in lytic infections occurring in transient aplastic anemia or non-immune hydrops fetalis by the detection of viral inclusion bodies in erythroid precursor cells. In all other PB19-associated diseases, molecular pathological methods must be applied.

Quantitative real-time polymerase chain reaction is used to determine the viral load in formalin-fixed and paraffin-embedded tissues derived from various organs. Using in situ hybridization it is demonstrated that endothelial cells of the microcirculatory periphery of the heart and hepatobiliar system in lytic infections are PB19-specific target cells in children and adults. Because treatment of lytic PB19 infection has been successfully applied, the pathologist should be alerted to include PB19 into the diagnostic spectrum of viral disease, especially in immunocompromised patients [1].

The spectrum of diseases linked to PB19 primarily involves infection in the healthy host manifested as erythema infectiosum, arthropathy, and hydrops fetalis, as well as a number of hematological consequences in susceptible patients [2].

Asymptomatic Infection

Most persons with parvovirus PB19 infection remain asymptomatic and those who are seropositive for the virus have no recollection of previous symptoms. In one study, 32 percent of household contacts of patients with acute parvovirus PB19 infection reported no symptoms at the time that they had parvovirus-specific IgM antibodies [3].

ERYTHEMA INFECTIOSUM (FIFTH DISEASE)

It is a childhood exanthem characterized by a "slapped cheek" rash. Fifth disease takes its name from a list of common childhood exanthems, named in the order of the dates when they were first reported: measles, scarlet fever, rubella, Duke's (or fourth) disease, fifth disease, and roseola, or sixth disease[4].

The classic course of erythema infectiosum can be divided into three distinct stages. The first stage (occurring after an incubation period of four to 14 days) is manifested by a mild prodromal illness characterized by low-grade fever, headache and gastrointestinal symptoms. This stage, which is often unrecognized, corresponds with the period of viremia and the period of contagion [3].

The second stage of the illness, occurring three to seven days after the prodrome, is characterized by the appearance of bright erythematous facial exanthems. Because this exanthem most commonly involves the malar eminences and spares the nasal bridge and perioral areas, the characteristic "slapped-cheek" appearance becomes evident. This stage is seen more commonly in children than in adults, and the exanthem may become more marked with exposure to sunlight [5].

The third stage of the illness occurs one to four days after the appearance of the facial exanthem and is characterized by the appearance of lacy, erythematous, maculopapular exanthems on the trunk and extremities. This eruption may be pruritic and often is evanescent, recurring over one to three weeks. Because the appearance of the exanthem corresponds with the development of antibody, patients with the rash of erythema infectiosum are no longer contagious. Although it is helpful to classify the stages of erythema infectiosum, the distinct features may be variable. For example, the facial exanthem may be pronounced in some patients but not in others. Similarly, the third stage of the illness may range from a very faint erythema to a florid confluent eruption [3].

During the past 2 decades, other unusual skin eruptions have been noted in association with PB19 infection. Papular purpuric gloves and socks syndrome is one form of parvovirus infection described mostly in white, young adult patients. Recently, there have been reports of a few patients with acropetechial syndrome consisting of a papular purpuric gloves and socks syndrome-like presentation with additional involvement of the perioral and chin area, temporally associated with acute PB19 infection [6].

Differential Diagnosis of Erythema Infectiosum

Erythema infectiosum must be differentiated from other causes of viral exanthems which causes centrally distributed maculopapular eruptions. These eruptions include rashes that begin centrally, first affecting the head and neck, and then progress peripherally, these include rubeola, rubella and roseola [7].

The exanthema of rubeola begins around the fourth febrile day, with discrete lesions that become confluent as they spread from the hairline downward, sparing the palms and soles. The exanthema typically lasts four to six days. The lesions fade gradually in order of appearance, leaving a residual yellow-tan coloration or faint desquamation. Rubeola is also distinguished by the presence of Koplik's spots in the oral mucosa [8].

Rubella is similar to rubeola. However, it causes less severe symptoms, and its exanthem characteristically has a shorter duration (two to three days)[8]. Roseola, or exanthema subitum, is caused by human herpesvirus 6. This disease occurs in children less than three years of age. As in fifth disease, the rash appears after the resolution of several days of high fever. The diffuse maculopapular eruption often spares the face and is of short duration, typically fading within three days [9].

Lyme Disease is an infection associated with skin eruptions. It is the most commonly reported vector-borne illness in the United States. It is caused by the spirochete Borrelia burgdorferi, which is transmitted by the bite of a tick (Ixodes species).

Erythema migrans, the pathognomonic rash, develops in about 80 percent of patients with Lyme disease.This enlarging, erythematous macular rash begins as a macule or papule at the site of inoculation. Systemic symptoms, including fever, chills, myalgias, headaches and arthralgias, often accompany the rash. The rash is more common on the proximal extremities, in body creases and on the chest. It enlarges over a period of days to weeks, reaching a maximum diameter of 3 to 68 cm (median diameter: 15 cm). The primary lesion may show central clearing, central necrosis, induration or vesiculation. Smaller secondary lesions may develop in up to 20 percent of patients with Lyme disease and may indicate early hematogenous spread [10].

Drug reactions can present as any dermatologic morphology and show no predilection for age, gender or race. Exanthematous eruptions most commonly occur in association with the administration of penicillins or cephalosporins. The rash usually appears within the first week after the offending drug is started and typically resolves within days after the drug is discontinued. Drug-related reactions can be difficult to distinguish from viral exanthems, but they may be more intensely erythematous and pruritic [8].

Gloves and Socks Syndrome

Parvovirus B19 has been associated with papular, purpuric gloves and socks syndrome, although a causative relationship has not been proven. The syndrome typically occurs in young adults and presents as symmetric, painful erythema and edema of the feet and hands. The condition gradually progresses to petechiae and purpura and may develop into vesicles and bullae with skin sloughing [11]. A hallmark of the syndrome is a sharp demarcation of the rash at the wrists and ankles, although other areas (e.g., cheeks, elbows, knees, inner thighs, glans penis, buttocks, or vulva) may be involved [12]. Most patients usually appear well but some patients may experience arthralgia, fever, or both. Symptoms usually resolve within one to three weeks without scarring. Gloves and socks syndrome also has been associated with hepatitis B, cytomegalovirus, Epstein-Barr virus, human herpesvirus 6, measles, coxsackievirus B, and drug reactions [13].

Arthropathy

In contrast to the mild course of the rash illness in children with PB19 infection, in adults, particularly middle-aged women, the infection may cause clinically significant arthropathy [14, 15]. Not only arthralgia but also inflammatory arthritis occur in about 50 percent of older patients; approximately 15 percent of new cases of arthritis may represent the sequelae of PB19 infection. Symmetric joint involvement, usually of the hands and occasionally of the ankles, knees, and wrists, can mimic rheumatoid arthritis, and the results of a test for PB19 usually resolves within a few weeks, and even when symptoms persist for months or years, joint destruction does not occur. The pathogenesis of PB19 arthropathy is assumed to involve deposition of immune complexes [4]. PB19 DNA may be present in inflamed joints [16].

Autoantibody and low complement were seen in acute human parvovirus infection, and parvovirus B19 infection present clinically lupus like tableau [17]. The joint abnormalities predominate in the hands and feet and usually resolve within a week or two (range 2-21 d). Serological tests show IgM antibodies against B19, confirming the diagnosis of recent infection.

Protracted polyarthritis occurs in some patients and seems associated with the DR4 histocompatibility alleles. Rheumatoid factors can be produced transiently in these patients. Other autoantibodies produced in the wake of PB19 infection include anti-nuclear antibodies, anti-DNA, anti-SSA/SSB, and anti-phospholipids. Acute B19 infection can simulate early rheumatoid arthritis (RA) or systemic lupus erythematosus (SLE) (lupus-like eruption over the cheeks, cytopenia, etc.) [18].

The role of PB19 in rheumatoid arthritis has not been proved, moreover; PB19 arthropathy does not progress to rheumatoid arthritis [19]. However, case reports suggest that parvovirus infection may mimic, precipitate, or exacerbate juvenile rheumatoid arthritis [20], systemic lupus erythematosus [21] and fibromyalgia [22].

Hydrops Fetalis and Stillbirth

Hydrops fetalis consists of an abnormal accumulation of fluid in two or more fetal compartments, including ascites, pleural effusion, pericardial effusion, and skin edema [23].

Hydrops fetalis is classified according to pathophysiology to immune hydrops that usually occur in10% of cases due to antibody-mediated hemolytic anemia, Rhesus blood group system incompatibility and ABO blood group incompatibility. Non-Immune Hydrops usually occur in 90% of cases due to primary heart failure that result from congenital structural heart defect, arrhythmia and congenital infective myocarditis. Congenital infections known to cause hydrops by various mechanisms include PB19, Toxoplasmosis, Syphilis, Cytomegalovirus, Rubella, Leptospirosis, and Chagas' disease. High-output cardiac failure also lead to hydropes fetalis that: occur with anemia as twin-twin transfusion, feto-maternal hemorrhage, hemoglobinopathy and congenital infections. Vascular malformation also lead to hydropes fetalis. Anomalies of both blood vessels and lymohatic vessels usually are seen in congenital malformation, congenital tumor & thrombosis. Increased capillary permeability that occurs with congenital infections is known as cause of hydropes fetalis. Another cause is ecreased plasma oncotic pressure as in congenital liver or renal disease & hepatitis [24].

Hydrops fetalis & stillbirth were first reported in association with PB19 infection in 1984[25]. Between one third and two thirds of pregnant women in different parts of the world are susceptible to PB19 infection.

The out come of infection usually depends on immune state of the mother. No vertical transmission has been described if the mother is immune at the time of exposure. It is highly unlikely that fetal infection occurs if the mother has IgG antibodies since this is thought to give life-long protection against re-infection with PB19. When maternal infection occurs, maternal viremia reaches its peak approximately one week after infection. Symptoms such as erythema infectiosum, mild fever, arthralgia and headache start approximately 10-14 days after infection in about 50% of infected women. At the time of the occurrence of IgM antibodies, presumably during the maternal peak viral load (day 7), the risk of vertical transmission may be maximal [26].

Acute infection by PB19 during pregnancy has been associated with fetal anemia, hydrops fetalis, non-hydropic intrauterine fetal death, and asymptomatic fetal infection [27]. PB19 infects the fetal liver, the site of erythrocyte production during early development. The swollen appearance in hydrops is the result of severe anemia and perhaps also myocarditis, both of which contribute to congestive heart failure [28].

Pathognic effects of Parvovirus vertical transmission from mother to foetus is illustrated in figure 4 [29,30].

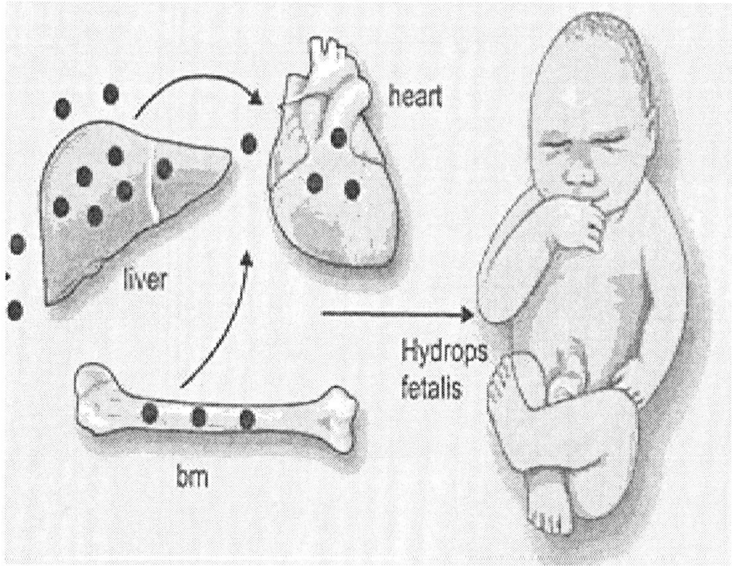

Figure 4. Vertical transmission of B19 from a primary infected mother may cause foetal infection.

So, vertical transmission of PB19 from a primary infected mother may cause foetal infection. Pathogenic mechanisms mainly include development of acute anemia upon infection of foetal haematopoietic cells. In early pregnancy haematopoiesis is seen in the liver and in later pregnancy this shifts to the bone marrow. The anemia may resolve spontaneously or proceed by causing cardiac failure and development of hydrops fetalis and in rare cases foetal death. The virus may also cause myocarditis and heart arrest by direct infection of myocardial tissue.

Thrombocytopenia may accompany severe anemia [31]. Seroprevalence data indicate that about half of pregnant women are susceptible to parvovirus infection [32]. Infection during the second trimester poses the greatest risk of hydrops fetalis. PB19 probably accounts for 10 to 20 percent of all cases of non-immune hydrops fetalis [33].

The risk of adverse fetal outcome is increased if maternal infection occurs during the first two trimesters of pregnancy but may also happen during the third trimester. Previous data demonstrated a decreasing expression of globoside within the villous trophoblast layer of the human placenta with increasing gestational age [34]. This may explain why the incidence of fetal morbidity and mortality related to PB19 infection decreases with gestational age [35]. More likely, passive transfer of maternal antibodies after the age of 25 weeks may reduce fetal morbidity and mortality at late stages of pregnancy.

It is a significant cause of fetal loss throughout pregnancy, but has a higher impact in the second half of pregnancy when spontaneous fetal loss from other causes is relatively rare. Parvovirus infection can cause severe fetal anemia as a result of fetal erythroid progenitor cells infection with shortened half life of erythrocytes, causing high output cardiac failure and therefore nonimmune hydrops fetalis (NIHF). The P antigen expressed on fetal cardiac myocytes enables the Parvovirus B19 to infect myocardial cells and produce myocarditis that aggravates the cardiac failure. Although there are several reports of major congenital anomalies among offspring of mothers infected by Parvovirus, the virus does not seem to be a significant teratogen. Since Parvovirus B19 infection can cause severe morbidity and mortality, it should be part of the routine work up of complicated pregnancies. Risk assessment for maternal infection during pregnancy is especially important during epidemics when sero-conversion rates are high [36].

Most PB19 infections during pregnancy do not lead to loss of the fetus. The few reported cases of developmental anomalies in the eyes or nervous system of infants with in utero exposure may be coincidental. Few cases of encephalopathy and severe CNS abnormalities following intrauterine B19V have been reported and only three cases of B19V neonatal encephalitis / meningitis have been reported [37,38]. The low frequency of neurological complications suggests that this is an uncommon complication. PB19 has also been associated with pediatric stroke [39, 40]. Isumi et al.[38] demonstrated perivascular calcifications in the fetal cerebralcortex, basal ganglia, thalamus and germinal layer following congenital PB19 infection. In, the cerebral white matter multinucleated reactive microglial cells were seen. PB19V DNA could be detected in glial cells and endothelial cells and suggest that immature fetal blood vessels permit infection of PB19V, leading to perivascular inflammatory changes.

Another reported complication of intrauterine infection with PB19 is severe anemia at birth with bone marrow morphologic features that are consistent with constitutional pure red-cell aplasia (Diamond–Blackfan anemia) or congenital dyserythropoietic anemia may be due to transplacental transmission of PB19 infection [41].

In the case of Parvovirus B19 exposure, it is advisable to control the maternal serology to know its initial status. According to the result, a weekly ultrasonographic supervision will be proposed to detect foetal anaemia (ascites, pericardial effusion). In the case of foetal hydrops, an in utero transfusion reduce the risk of foetal loss. The long-term outcome of infected foetuses is mostly good [42].

An intrauterine therapy with packed red cells could be performed for hydrops fetalis and low haemoglobin concentration. Investigation for the development and clinical testing of an efficient vaccine against parvovirus B 19 is currently in progress [43].

Meningitis and Neurological Disease

Prior to the discovery that PB19 was the etiological agent of erythema infectiosum, 2 cases of encephalitis associated with fifth disease were reported [44]. In both, encephalitis followed the appearance of a classical fifth disease rash. There was coincident fifth disease in the community at the time, and, despite extensive evaluation, no other cause for the encephalitis was found. In the first case the 8 year-old child made a full recovery, but in the second, there were permanent neurological sequelae. Both encephalitis and more often aseptic meningitis have been described in serologically confirmed PB19 infection [45], with detection of PB19 DNA in cerebrospinal fluid. In all these cases there has been no long-term neurological squeal.

Recently, PB19 has been suggested to have a role in acute cerebellar ataxia. A 2 year old boy developed acute cerebellar ataxia in association with erythema infectiosum. During the disease, genomic DNA and antibodies against human PB19 were detected in serum but not in cerebrospinal fluid. PB19 associated acute cerebellar ataxia might occur due to transient vascular reaction in the cerebellum during infection [46].

Other neurological symptoms have also been related to PB19 infection. Brachial plexus neuropathy with weakness and sensory loss has also been described in patients with PB19 infection [4], and in one study 50% of patients with classical fifth disease (confirmed serologically) experienced neurological symptoms (tingling and numbness in the fingers or toes)[37]. In most of these patients neurological examination was normal apart from decreased sensation to light touch, but one patient developed more significant disease with progressive weakness of one arm. The mechanism for the neurological symptoms is unknown, but the earlier appearance of the rash suggests that the neuropathy may be immune-mediated [47].

Myocarditis

Human parvovirus B19 has been linked to a variety of cardiac diseases. A causal association between viral infection and cardiac disease was frequently postulated following the detection of B19 DNA by PCR in endomyocardial biopsy specimens. Several authors have postulated a causative role of B19 in cardiac disease, such as acute myocarditis [48-50], dilative cardiomyopathy or idiopathic left ventricular dysfunction [51, 52], and peripartum cardiomyopathy [53, 54].

There have been two fatal cases of PB19-associated myocarditis reported. One year-old child, developed cardiac insufficiency following serologically confirmed erythema infectiosum. There was a temporary response to digoxin and diuretics, but the child died two weeks later. On autopsy there was an active myocarditis with a mononuclear inflammatory infiltrate and severe myocyte necrosis. PB19 capsid proteins were detected in myocardial tissue sections. The myocardial findings were similar to those seen in fetuses infected in utero by PB19 [55]. In a second case, a 3 year-old child died of myocarditis following PB19 infection. Although PB19 could be detected in liver and spleen tissues, no PB19 DNA was detected in the myocardial tissue. The role of PB19 in the pathogenesis of myocarditis needs further investigations, particularly as P antigen is found on fetal myocardial cells and PB19 appears to cause myocarditis in the fetus [4].

PB19-associated inflammatory cardiomyopathy is characterized by infection of intracardiac endothelial cells of small arterioles and veins, which may be associated with endothelial dysfunction, impairment of myocardial microcirculation, penetration of inflammatory cells, and secondary myocyte necrosis. Recent observations showed that B19 is involved in intracellular calcium regulation by the viral phospholipase. B19-induced caspase activation can lead to proinflammatory/proapoptotic processes through dysregulation of STAT signaling. These cellular interactions may contribute to mechanisms by which B19 establishes persistent infection in endothelial cells and play a critical role in viral pathogenesis of inflammatory cardiomyopathy [56].

HEPATITIS & VASCULITIS

The role of PB19 in both hepatitis and vasculitis remains unclear. Although transient elevation of liver transaminases is not uncommon in PB19 infection, frank hepatitis associated with PB19 infection has rarely been reported [57]. PB19 has been suggested as a possible causative agent of fulminant liver failure and associated aplastic anemia based on PCR studies [58]. Parvovirus B19 induced acute hepatitis and hepatic failure has been previously reported, mainly in children. Very few cases of parvovirus induced hepatic failure have been reported in adults and fewer still have required liver transplantation [59].

Parvovirus B19 might be a cause of hepatic dysfunction by mechanisms that are not yet elucidated. Presumably, PB19 can bind to the cell surface containing globoside, the putative PB19 receptor, and thus penetrate liver cells [60,61]. Parvovirus B19 might be a cause of hepatic dysfunction by mechanisms that are not yet elucidated. Presumably, PB19 can bind to the cell surface containing globoside, the putative B19 receptor, and thus penetrate liver cells [60,61]. PB19 infection. Viral infection might induce increased levels of cytokines such as interferon gamma and tumour necrosis factor a that deregulate the phagocytic system, leading to pancytopenia and/or hepatic dysfunction [62-65]. In cases of unexplained hepatitis, PCR testing for HPV PB19 might help to clarify the aetiology.

Several case reports have described positive PB19 serology in patients with vasculitis and/or polyarteritis nodosum; however, in each individual report it was uncertain as to whether the association was coincidental or causative. More recently, PB19 infection has been associated with acute systemic necrotizing vasculitis. Recent infection with PB19 was indicated in 3 patients by the presence of both PB19 IgM in the serum and PB19 DNA in serum and tissues. Treatment with intravenous immunoglobulin led to clearing of the virus and resolution of the patients' symptoms [4].

Systemic Sclerosis

Systemic sclerosis is a multisystem disease that affects the skin and internal organs, including the gastrointestinal tract, lung, heart, kidney, and peripheral nervous system.240 B19 DNA was detected in the bone marrow of 12 of 21 (57%) patients with systemic sclerosis compared with 0 of 15 healthy controls ($p<0.01$); serum NS1 antibodies occurred in 33% of patients with systemic sclerosis compared with 13% of controls [66].

The spectrum of disease caused by parvovirus B19 has been expanding in recent years because of improved and more sensitive methods of detection. There is evidence to suggest that chronic infection occurs in patients who are not detectably immunosuppressed.

Chronic Fatigue Syndrome

It is suggested that how often chronic fatigue syndrome (CFS) appears after human parvovirus B19 (B19) infection and prolonged B19 viremia or some other factors cause CFS. Recent study suggests that we should consider the possibility of CFS after B19 infection [67]. Earlier data has demonstrated that circulating TNF-α and IFN-γ levels are raised during acute and convalescent parvovirus B19 infection and are associated with prolonged and chronic fatigue. These findings shed new light on our knowledge of the host response to this infection, its chronicity in certain individuals and the pathogenesis of B19 virus-associated fatigue. As B19 virus infects most people worldwide, these findings suggest that cytokines produced during the years following acute infection in certain individuals may predispose to the development of other diseases and provide clues as to the mechanism of other observed phenomena [68].

Uveitis

Two cases of B19 associated uveitis have been reported: one in an adult with EI, tonic pupils, and ophthalmoplegia, and one in a young girl associated with transient antinuclear antibody and rheumatoid factor production[69] Anti-PB19 IgG was found in vitreous fluid from three of six IgG seropositive cases of chronic intermediate uveitis, but none had evidence of local anti-B19 IgG production [70].

Myositis

Increases in muscle enzymes are occasionally seen in acute PB19 infection [71] and one child with new onset of juvenile dermatomyositis was shown to have acute PB19 infection. A childhood case of B19 associated interstitial lung disease, hepatitis, and myositis has been reported. [72]. In addition, B19 DNA was isolated from the skeletal muscle of an adult patient with mixed connective tissue disease

B) Hematological Consequences of PB19 Infection in Susceptible Patients[2]

Various hematological consequences are associated with PB19 infection in susceptible individuals with hemolytic anemia's. These complications include, transient Aplastic Crisis (TAC) in cases of haemolytic anemia, idiopathic thrombocytopenic purpura (ITP), neutropenia., congenital anemia, haemophagocytic syndrome, chronic bone marrow failure due to persistent PB19 infection. These complications will be discussed later on.

Prevention of PB19 Infection

Prevention of disease by isolating susceptible individuals is impractical since infections may be subclinical and symptomatic individuals are infectious before any sign of illness. The rash or arthralgia stages of PB19 infection are caused by immune reactions, not viremia and therefore not infectious. Theoretically, susceptible individuals with chronic haemolytic anemia or immunocompromised children could be temporarily protected by the administration of human immunoglobulin.

Recombinant PB19 virus-like particles can induce neutralizing antibodies and provide a potentially suitable method for vaccination of humans [79]. several studies [73-77] (including a randomized, double - blind, phase 1 trial) have reported on a recombinant human parvovirus B19 vaccine composed of the VP1 and VP2 capsid proteins and formulated with MF59C.1. All volunteers became parvovirus B19 seropositive after receiving at least 2 doses of vaccine. There were no deaths or serious adverse events during the study [76].

In September 2007 in the United States, the National Institute of Allergy and Infectious Diseases initiated a phase I/II study of the safety and immunogenicity of a recombinant human parvovirus B19 vaccine [77]. Perhaps, in the near future, the vaccine could be used to prevent transient aplastic crisis in patients with sickle cell disease or other hemolytic anemias and in immunodeficient patients with pure red cell anemia [78] and to prevent persistent arthropathy in adults or in parvovirus B19–seronegative women with hydrops fetalis (when inoculated early during pregnancy).

Treatment and Control of PB19 Infection

Most cases of PB19 infections are mild and self limiting and specific treatment is not required. However, there are situations in which sever anemia occurs as a consequence of PB19 infection, blood transfusion is indicated. In cases of aplastic crisis; the transfusion corrects the patient over a relatively short period of erythroid aplasia before the immune response rapidly clears the virus infection. In immunocompromised patients; there is a failure to produce neutralizing antibodies, so treatment consists of transfusion plus the administration of human normal immunoglobulin for 10 days. This leads to disappearance of the viremia, sometimes permanently and sometimes only temporarily [78].

Immunocompetent patients with TAC have a self-limiting illness, and typical transient aplastic crisis is readily treated by blood transfusion and supportive therapy alone. In patients with documented persistent B19 infection, administration of immunoglobulin (IVIG) can be beneficial [79, 80]. The current recommendation is IVIG, 0.4 g/kg for 5 days.

Patients often respond with a marked reduction in the level of PB19 viraemia, reticulo-cytosis and resolution of the anemia within 1-2weeks of treatment.However, patients should be monitored for evidence of relapse, by observation of the reticulocyte counts, and assays for B19 viraemia when indicated. If relapse occurs less than 6 months after the initial treatment, especially in HIV positive patients, an empirical maintenance treatment with a single-day infusion of 0.4 g/kg IVIG every 4 weeks may control the PB19 viraemia.

On the other hand, patients with PB19 induced arthralgia may need symptomatic treatment as anti-inflammatory drugs [2].

PB19 infection in pregnant seronegative women should be monitored by weekly ultrasound examination. In cases of hydrops fetalis, cordocentesis and intrauterine transfusion are effective in lowering the mortality of the fetus [81]

Generations of neutralizing human monoclonal antibodies against PB19 proteins have been proposed as an immunotherapy of clinically infected individuals and actually infected pregnant women.

Conclusion

The spectrum of diseases linked to PB19 primarily involves infection in the healthy host manifested as erythema infectiosum, gloves and stokes syndrome, arthropathy, and hydrops fetalis, as well as a number of hematological consequences in susceptible patients. This virus has been linked to other clinical syndrome as mycodarditis, vasculitis and meningitis. Treatment depends mainly on specific immunoglobulins as indicated. Trials for vaccine developments are undertaken.

References

[1] Bultmnn BD, Klingel K, Sotlar K, et al. Parvovirus B19 a pathogen responsible for more than hematological disorder. *Virchows Arch.* 2003, 442: 8-17.

[2] Heegaard ED and Brown KE. Human Parvovirus B19. *Clin. Microb. Rev.* 2002, 15:485-505.

[3] Broliden K., Tolfvenstam T & Norbeck O. Clinical aspects of parvovirus B19 infection. *J. of Int. Med.* 2006; 260: 285–304

[4] Young NS and Brown KE. Parvovirus B19. *NEJM.* 2004,350:586-97.

[5] Weir E. Parvovirus B19 infection: fifth disease and more. *Canad. Med. Ass. J.* 2005, 172: 743.

[6] McNeely M, Friedman J and Pope E. Generalized petechial eruption induced by parvovirus B19 infection. *J. Am. Acad. Dermatol.* 2005, 52: 109-13.

[7] Weber DI, Cohen MS and Fine JD. The acutely ill patient with fever and rash. In: Mandell GL, Bennett JE, Dolin R, eds. Mandell, Douglas, and Bennett's Principles and practice of infectious diseases. 1999, 5th ed. Philadelphia: Churchill Livingstone.p:633-50.

[8] Kaye ET and Kaye KM Fever and rash. In: Fauci AS, et al. eds. Harrison's Principles of internal medicine. 1998, 14th ed. New York: McGraw-Hill, Health Professions Division.

[9] Nelson WE, Behrman RE, Kliegman RM, et al. eds. Nelson Textbook of pediatrics. 1996, 15th ed. Philadelphia: Saunders.

[10] Harry D and Howard T. Evaluating the Febrile Patient with Rash. American Family Physicians. 2000, On web site: http://www.findarticles.com/p/articles/mi_m3225/is_4_62/ai_65106197

[11] Alfadley A, Aljubran A, Hainau B, Alhokail A. Papular-purpuric "gloves and socks" syndrome in a mother and daughter. *J. Am. Acad. Dermatol.* 2003; 48:941-4.

[12] Metry D, Katta R. New and emerging pediatric infections. *Dermatol. Clin.* 2003; 21:269-76.

[13] Katta R. Parvovirus B19. a review. *Dermatol. Clin.* 2002; 20:333-42.

[14] Moore TL. Parvovirus-associated arthritis. *Curr. Opin. Rheumatol.* 2000, 12:289-94.

[15] Kerr JR. Pathogenesis of human parvovirus B19 in rheumatic disease. *Ann. Rheum. Dis.* 2000,59:672-83.

[16] Söderlund M, von Essen R, Haapasaari J, et al. Persistence of parvovirus B19 DNA in synovial membranes of young patients with and without chronic arthropathy. *Lancet.* 1997,349:1063-65.

[17] Kumano K. Various clinical symptoms in human parvovirus B19 infection. *Nihon Rinsho Meneki Gakkai Kaishi.* 2008; 31:6/448-53.

[18] Meyer O. Parvovirus B19 and autoimmune diseases. *Joint Bone Spine.* 2003;70 (1):6-11

[19] Speyer I, Breedveld FC and Dijkmans BAC. Human parvovirus B19 infection is not followed by inflammatory joint disease during long term follow-up: a retrospective study of 54 patients. *Clin. Exp. Rheumatol.* 1998, 16:576-78.

[20] Lehmann HW, von Landenberg P and Modrow S. Parvovirus B19 infection and autoimmune disease. *Autoimmun. Rev.* 2003,2:218-23.

[21] Tanaka A, Sugawara A, Sawai K, et al. Human parvovirus B19 infection resembling systemic lupus erythematosus. *Intern. Med.* 1998, 37:708-10.

[22] Leventhal NJ, Naides SJ and Freundlich B Fibromyalgia and parvovirus infection. *Arthritis Rheum.* 1991,34:1319-24.

[23] Skoll MA, Sharland GK and Allan LD. Is the ultrasound definition of fluid collections in non-immune hydrops fetalis helpful in defining the underlying cause or predicting outcome? *Ultrasound Obstet Gynecol.* 1991, 1:309-12.

[24] Behrman RE, Kliegman RM and Nelson HBJ. Textbook of Pediatrics, 16th edition. 2000, Philadelphia: W.B. Saunders Company.

[25] Anderson LJ. Role of Parvovirus B19 in Human Disease. *Pediatr. Infect. Dis. J.* 1987,6:711-18.

[26] De Haan TR, Oepkes D, Beersma MFC, Walther FJ. Etiology, diagnosis and treatment of hydrops foetalis. *Cur. Pediat. Rev.* 2005; 1: 63-72.

[27] Enders M, Weidner A, Zoellner I, et al. Fetal Morbidity and Mortality after Acute Human Parvovirus B19 Infection in Pregnancy: Prospective Evaluation of 1018 Cases. *Obstet. & Gynecol. Surv.* 2005, 60(2):83-84.

[28] Morey AL, Keeling JW, Porter HJ, et al. Clinical and histopathological features of parvovirus B19 infection in the human fetus. *Br. J. Obstet. Gynaecol.* 1992, 99:566-74.

[29] Broliden K., Tolfvenstam T. & Norbeck O. Clinical aspects of parvovirus B19 infection. *J. of Int. Med.* 2006; 260: 285–304

[30] Anderson LJ, Young NS (eds). Human Parvovirus B.19, Vol. 20. Basel: Karger, 1997

[31] Forestier F, Tissot JD, Vial Y, et al. Hematological parameters of parvovirus B19 infection in 13 fetuses with hydrops foetalis. *Br. J. Haematol.* 1999,104:925-27.

[32] Miller E, Fairley CK, Cohen BJ, et al. Immediate and long term outcome of human parvovirus B19 infection in pregnancy. *Br. J. Obstet. Gynaecol.* 1998, 105:174-78.

[33] Jordan J, Tiangco B, Kiss J, et al. Human parvovirus B19: prevalence of viral DNA in volunteer blood donors and clinical outcomes of transfusion recipients. *Vox. Sang.* 1998, 75:97-102.

[34] Jordan JA, DeLoia JA. Globoside expression within the human placenta. *Placenta.* 1999; 20: 103-8.

[35] Enders M, Weidner A, Zoellner I, Searle K, Enders G. Fetal morbidity and mortality after acute human parvovirus B19 infection in pregnancy: prospective evaluation of 1018 cases. *Prenat. Diagn.* 2004; 24: 513-8.

[36] Ergaz Z, Ornoy A. Parvovirus B19 in pregnancy. *Reprod. Toxicol.* 2006; 21:4/421-35.

[37] Kerr JR, Barah F, Chiswick ML, McDonnell GV, Smith J, Chapman MD, Bingham JB,Kelleher P, Sheppard MN. Evidence for the role of demyelination, HLA-DR alleles, and cytokines in the pathogenesis of parvovirus B19 meningoencephalitis and its sequelae. *J. Neurol. Neurosurg. Psychiatry.* 2002; 73:739-46.

[38] Isumi H, Nunoue T, Nishida A, Takashima S. Fetal brain infection with human parvovirus B19. *Pediatr. Neurol.* 1999; 21: 661-3.

[39] Guidi B, Bergonzini P, Crisi G, Frigieri G, Portolani M. Case of stroke in a 7-year-old male after parvovirus B19 infection. *Pediatr. Neurol.* .2003; 28:69-71.

[40] Craze JL, Salisbury AJ, Pike MG. Prenatal stroke associated with maternal parvovirus infection. *Dev. Med. Child Neurol.* 1996; 38:84-5.

[41] de Haan T.R., Wezel-Meijler G., Beersma v. von Lindern, J.S,. van Duinen S.G, Walther F.J., Fetal stroke and congenital Parvovirus B19 infection complicated by activated protein C resistance. *Acta Paediatrica.* 2006; 95: 863-867

[42] Brown KE, Green SW, Antunez de Mayolo J, et al. Congenital anemia after transplacental B19 parvovirus infection. *Lancet.* 1994, 343:895-96.

[43] Brochot C, Debever P, Subtil D, Puech F. Which supervision and treatment to use in case of exposure to Parvovirus B19 during pregnancy?] *Gynecol. Obstet. Fertil.* 2008 ;36:2/204-11.

[44] Mylonas I, Gutsche S, Anton G, Jeschke U, Weissenbacher ER, Friese K. Parvovirus B 19 infection during pregnancy. *Z. Geburtshilfe Neonatol.* 2007; 211:2/60-8.

[45] Hall CB and Horner FA. Encephalopathy with erythema infectiosum. *Am. J. Dis. Child.* 1997,131:65-7.

[46] Suzuki N, Terada S and Inoue M. Neonatal meningitis with human parvovirus B19 infection [letter]. *Arch. Dis. Child.* 1995, 73:196-97.

[47] Shimizu Y, Ueno T, Komatsu H, et al. Acute cerebellar ataxia with human parvovirus B19 infection. *Arch. Dis. Child.* 1999,80:72-73

[48] Faden H, Gary GW and Korman M. Numbness and tingling of fingers associated with parvovirus B19 infection [letter]. *J. Infect. Dis.* 1990, 161:354-55.

[49] Schowengerdt K O, Ni J, Denfield S W, Gajarski R J, BowlesN E, Rosenthal G, Kearney D L, Price J K, et al. Association of parvovirus B19 genome in children with myocarditis and cardiac allograft rejection:

diagnosis using the polymerase chain reaction. *Circulation.* 1997, 963549-3554.

[50] Dettmeyer, R., R. Kandolf, A. Baasner, S. Banaschak, A. M. Eis-Hubinger, and B. Madea. Fatal parvovirus B19 myocarditis in an 8-year-old boy. *J. Forensic. Sci.* 2003, 48183-186

[51] Kühl U, Pauschinger M, Seeberg B, Lassner D, Noutsias M, Poller W, Schultheiss H P. Viral persistence in the myocardium is associated with progressive cardiac dysfunction. *Circulation.* 2005, 112:1965-1970

[52] Donoso Mantke O, Nitsche A, Meyer R, Klingel K, Niedrig M. Analysing myocardial tissue from explanted hearts of heart transplant recipients and multi-organ donors for the presence of parvovirus B19 DNA. *J. Clin. Virol.* 2004, 3132-39.

[53] Klein R M, Jiang H, Niederacher D, Adams O, Du M, Horlitz M, P et al. Frequency and quantity of the parvovirus B19 genome in endomyocardial biopsies from patients with suspected myocarditis or idiopathic left ventricular dysfunction. *Z. Kardiol.* 2004, 93300-309

[54] Bultmann B D, Klingel K, Nabauer M, Wallwiener D, Kandolf R. High prevalence of viral genomes and inflammation in peripartum cardiomyopathy. *Am. J. Obstet. Gynecol.* 2005, 193363-365.

[55] Schenk T, Enders M, Pollak S, Hahn R, Huzly D. High prevalence of human parvovirus B19 DNA in myocardial autopsy samples from subjects without myocarditis or dilative cardiomyopathy. *J. Clin. Microbiol.* 2009, 47,1:106-110.

[56] Saint-Martin J, Choulot JJ, Bonnaud E, et al. Myocarditis caused by parvovirus [letter; comment]. *J. Pediatr.* 1990,116:1007-8.

[57] Kandolf R, Bültmann B, Klingel K, Bock CT. Molecular mechanisms and consequences of cardiac viral infections. *Pathologe.* 2008; 29 Suppl 2:112-7.

[58] Yoto Y, Kudoh T, Haseyama K, et al. Human parvovirus B19 infection associated with acute hepatitis. *Lancet.* 1996, 347:868-69.

[59] Langnas AN, Markin RS, Cattral MS, et al. Parvovirus B19 as a possible causative agent of fulminant liver failure and associated aplastic anemia. *Hepatology.* 1995,22:1661-65.

[60] Krygier DS, Steinbrecher UP, Petric M, Erb SR, Chung SW, Scudamore CH, Buczkowski AK, Yoshida EM. Parvovirus B19 induced hepatic failure in an adult requiring liver transplantation. *World J. Gastroenterol.* 2009, 28; 15(32):4067-9

[61] Karetnyi YV, Beck PR, Markin RS, Langnas AN, Naides SJ. Human parvovirus B19 infection in acute fulminant liver failure. *Arch. Virol.* 1999; 144:1713–24.

[62] Pinho JR, Alves VA, Vieira AF, et al. Detection of human parvovirus B19 in a patient with hepatitis. *Braz. J. Med. Biol. Res.* 2001; 34:1131–8.

[63] Smith MA, Ryan ME. Parvovirus infections. From benign to life-threatening. *Postgrad. Med.* 1988; 84:127–38.

[64] Muir K, Todd WT, Watson WH, Fitzsimons E. Viral-associated haemophagocytosis with parvovirus-B19-related pancytopenia. *Lancet.* 1992; 339:1139–40.

[65] Ohga S, Matsuzaki A, Nishizaki M, et al. Inflammatory cytokines in virus-associated hemophagocytic syndrome. Interferon-gamma as a sensitive indicator of disease activity. *Am. J. Pediatr. Hematol. Oncol.* 1993; 15:291–8.

[66] Kenji S, Hiroyuki T, Parvovirus B19-associated haemophagocytic syndrome in healthy adults. *Br. J. of Haematol.* 1995, 89(4):923-926,

[67] Ferri C, Zakrzewska K, Longombardo G, Giuggioli D, Storino FAA, Pasero G, et al. Parvovirus B19 infection of bone marrow in systemic sclerosis patients. *Clin. Exp. Rheumatol.* 1999; 17:718–20.

[68] Seishima M, Mizutani Y, Shibuya Y, Arakawa C Chronic fatigue syndrome after human parvovirus B19 infection without persistent viremia. *Dermatology.* 2008; 216(4):341-6.

[69] Kerr JR, Barah F, Mattey DL, et al. Serum tumour necrosis factor-α (TNF-α) and interferon-γ (IFN-γ) are detectable during acute and convalescent parvovirus B19 infection and are associated with prolonged and chronic fatigue. *J. Gen. Virol.* 2001, 82:3011–19.

[70] Maini RN, Edelstein C. Uveitis associated with parvovirus infection. *Br. J. Ophthalmol.* 1999; 83:1403–4.

[71] De Boer JH, de Keizer RJW, Kijlstra A. In search of intraocular antibody production of parvovirus B19 and adenovirus in intermediate uveitis. *Br. J. Ophthalmol.* 1993; 77:829.

[72] Lewkonia RM, Horne D, Dawood MR. Juvenile dermatomyositis in a child infected with human parvovirus B19. *Clin. Infect. Dis.* 1995; 21:430–2.

[73] Bousvaros A, Sundel R, Thorne GM, McIntosh K, Cohen M, Erdman DD, et al. Parvovirus B19-associated interstitial lung disease, hepatitis and myositis. *Pediatr. Pulmonol.* 1998; 26:365–9.

[74] Bansal GP, Hatfield JA, Dunn FE, et al. Candidate recombinant vaccine for human B19 parvovirus. *Infect. Dis.* 1993; 5:1034–44
[75] Tsao EI, Mason MR, Cacciuttolo MA, et al. P of parvovirus B19 vaccine in insect cells co□infected with double baculoviruses *Biotechnol. Bioengineer.* 2000, 49:130-8
[76] MF59c.1-adjuvanted recombinant vaccine reported safe, immunogenic. Vaccine Wkly 2003. Available athttp://www.newsrx.com/article.php?articleID=135276
[77] Ballou R, Reed J, Noble W, Young N, Koenig S. Safety and immunogenicity of a recombinant parvovirus B19 vaccine formulated with MF59C.1. *J. Infect. Dis.* 2003; 187:675–8.
[78] Gigler A, Dorsch S, Hemauer A, et al. Generation of neutralizing human monoclonal antibodies against parvovirus B19 proteins. *J. Virol.* 1999,73:1974-79
[79] Cubie HA. Viral pathogens an associated diseases (Parvoviruses) in Medical Microbiology.16th edition. Greenwood - Churchill Livingstone. 2002, 448-54.
[80] Frickhofen N, Abkowitz JL, Safford M et al. Persistent B19 parvovirus infection in patients infected with human immunodeficiency virus type-1 (HIV-1): a treatable cause of anemia in AIDS. *An. of Int. Med.* 1990; 113: 926-933.
[81] Kurtzman GJ, Cohen B, Meyers P et al. Persistent B19 parvovirus infection as a cause of severe chronic anaemia in children with acute lymphocytic leukaemia. *Lancet.* 1988; ii: 1159-1162.
[82] Schild RL, Bald R, Plath H, et al. Intrauterine management of fetal parvovirus B19 infection. *Ultrasound Obstet. Gynecol.* 1999,13:161-66.

Chapter 3

PB19 Infection in Immunodeficiency Disorders

Abstract

Acute parvovirus B19 infection produces erythema infectiosum in normal individuals or even pass as asymptomatic infection, but in immunocompromised patients with insufficient production of neutralizing antibody response due to persistent bone marrow insufficiency the infection may cause wide spectrum of clinical conditions varying from pure red cell aplasia, chronic anaemia, to transient aplastic crisis in patients with hemolytic anaemia. Predisposing conditions include congenital immunodefeceincy syndrome, acute lymphoblastic leukaemia, large granular lymphocyte leukaemia, acute and chronic myeloid leukaemia. Moreover, Parvovirus B19 may be an important opportunistic pathogen in all organ transplant recipients of heart, bone marrow, liver and kidneys and patients with congenital or acquired immunodeficiency syndromes including (AIDS), lymphoproliferative disorders and other malignancies.

Keywords: PB19 infection, immunocompromized patients.

Introduction

Persistent infection with PB19 is frequently encountered in patients with various immunodeficiency disorders.

The common immunodeficiency disorders that have been associated with chronic PB19 infection are summarized in table 2 [1, 2].

Table 2. The common immunodeficiency disorders that have been associated with chronic PB19 infection [1, 2]

Congenital Immunodeficiency: - Nazelof's syndrome. - Severe combined immunodeficiency. - Common variable immunodeficiency. Acquired Immunodeficiency: - HIV. - Malignancy: - Acute lymphoplastic leukaemia (ALL). - Acute myeloid leukaemia (AML). - Non-Hodgkin's lymphoma. - Brain tumors. - Wilms tumor. - Rabdomyosarcoma. - Organ transplantation: - Renal transplantation. - Cardiac transplantation - Liver transplantation. - BM transplant recipient. - Collagen vascular diseases: -Systemic Lupus Erythematosis. - Reumatoid arthritis.

In the absence of a neutralizing antibody response, persistent B19 infection results, generally manifesting as pure red cell aplasia (PRCA). PRCA can occur in a wide variety of immunosuppressed states: congenital immunodeficiency [3], acquired immunodeficiency (HIV infection or AIDS)[4], lymphoproliferative disorders[5], and after organ transplant. Immunosuppression may be subtle, and PRCA may be the presenting illness of patients with HIV infection or with very limited immune dysfunction [6]. In all cases, the stereotypical presentation is with persistent anemia associated with reticulocytopenia. Other blood lineages may also be affected, a case of recurrent agranulocytosis ascribed to persistent parvovirus B19 has been described [7] and in rare cases patients may present with pancytopenia. Examination of the bone marrow generally reveals the presence of scattered giant pronormoblasts.

There is persistent or recurrent parvoviraemia as detected by the presence of PB19 DNA in the serum, and anti-PB19 specific antibody is absent or low, with antibody to VP2 proteins, but not to VP1 [6]. Administration of neutralizing antibody in the form of intravenous immunoglobulin (IVIG) can be beneficial and ameliorative and sometimes curative.

Temporary cessation of maintenance chemotherapy may also lead to resolution of anemia, suggesting that decreasing the level of iatrogenic immunosuppression may allow the host to produce antibody and resolve the virus infection [8].The prevalence of PB19 induced anemia in HIV sero-positive patients is probably higher than recognized at present. In one early study of 50 patients with AIDS, no patients with PB19 viraemia were identified. In a larger cohort study PB19 DNA was found in only 1 of 191 (0.5%) HIV-seropositive homosexuals. However, PB19 DNA was found in 4 of 24 (17%) transfusion dependent HIV-seropositive homosexuals, and when a haematocrit of <22 was used as a criterion, 4/17 (24%) were positive [9]. In contrast to the earlier studies, the marrow morphology need not be suggestive of PB19 infection and giant pronormoblasts may not be present.

Immunodeficient Patients

Congenital Immunodeficiency Syndrome

Persistent PB19 infection producing PRCA has been reported in several patients with proven or suspected congenital immunodeficiency syndromes. Failure of the humoral immune response to clear parvovirus B19 in such patients results in persistent pure red cell aplasia[10]. There were various case reports about PB19 infection in congenital immunodefeceincy syndrome. The first report was of a 27-month-old boy with Nezelof's syndrome (a combined T and B cell defect in which serum immunoglobulin (Ig) is present but at low levels). This child presented with abrupt onset anemia and reticulocytopenia, as well as neutropenia. High concentrations of PB19 parvovirus (~10^8 genome copies/ml) were found in the serum coincident with the onset of anemia and persisted for five months. Giant pronormoblasts were present on marrow smears, and PB19 virus was detectable by molecular methods in bone marrow cells. The patient was treated with red cell transfusions and commercial immunoglobulin preparations. Remission of the anemia was associated with absence of virus in the serum and relapse with its reappearance [2, 3]. Similar findings have been described in a 21-month-old boy with common variable immunodeficiency (predominantly CD4 lymphopenia, with normal immune-globulin levels) [11] and a 12-month-old girl with severe combined immunodeficiency [12].

The most dramatic example of chronic marrow failure due to PB19 infection occurred in a 24-year-old male, who presented with a 10-year history of PRCA. PB19 DNA was found in serum and bone marrow cells at presentation. His sibling was remarkably had also developed PRCA simultaneously, but died later of complications secondary to transfusion hemosiderosis. His brother also suffered from chronic PB19 infection demonstrated by post-mortem detection of PB19 DNA in his fixed spleen tissue [13].

On treatment with commercial immunoglobulins, the virus disappeared from the patient's bone marrow and serum, and the hemoglobin rose; once virus was no longer detectable in the blood, periodic Ig injections were discontinued and blood counts have remained normal for several years. Although there was no clinical history of increased susceptibility to infections, underlying immunodeficiency was suggested by skin test anergy, poor in vitro lymphocyte responses to mitogens, a very high helper/suppressor T cell ratio, low natural killer cell numbers and depressed serum immunoglobulin levels [13].

Acquired Immunodeficiency Syndrome (AIDS)

PRCA due to PB19 also occurs in the AIDS patients, sometimes as the first manifestation of infection with HIV [2].

In AIDS patients, persistent infection may result in chronic anemia, this anemia is not as severe as that resulting from TAC in sickle-cell disease but may last months or even years, and be confused with other anemia's that may complicate the course of AIDS. The clinical picture is one of bone marrow failure or hypoplasia and is usually restricted to the erythrocytic lineage and responded to immunoglobulin therapy. In contrast, in HIV-positive patients without AIDS the infection evolves as a mild exanthemata disease or may be entirely asymptomatic and need no treatment. Explanations of persistent anemia due to PB19 infection in AIDS patients are plentiful, but there are fewer accounts of exanthematous disease or asymptomatic infection in HIV positive patients without AIDS [14].

The persistence of PB19 infection in immunocompromised patients has been reported as a cause of chronic hyporegenerative anemia [15]. Up to 25% of severe chronic anemia in AIDS has been ascribed to PB19 infection.

Knowing that an AIDS patient has chronic PB19 anemia lessens concern about drug anemia, protects the patient from invasive diagnostic maneuvers, and prevents the patient from disseminating the infection. In AIDS patients with pure red cell aplasia, a search for PV-B19 DNA in the serum or in the BM is warranted.

In AIDS patients, persisting PB19 infection resulting from impaired immunity may, as it approaches or exceeds the normal red cell life span, result in anemia. This anemia is not as severe as that resulting from transient aplastic crisis in sickle cell disease but may last months or even years and be confused with other anemias that may complicate the course of AIDS. The clinical picture is one of bone marrow (BM) failure or hypoplasia and is usually restricted to the erythrocytic lineage (acquired pure red cell aplasia).

Exposure to community outbreaks of PB19, often seen in spring and autumn, should raise the possibility of PB19 infection in the differential diagnosis of diffuse rashes and arthralgia in adults with HIV infection [16]. In HIV-positive patients without AIDS the infection evolves as a mild exanthematous disease.

The diagnosis is established when the following criteria are met: (1) BM biopsy showing PRCA, (2) Serum or BM positive for PB19 DNA by PCR or dot-blot hybridization, and (3) No alternate explanation for the PRCA. Serological methods are unreliable for the diagnosis because these patients often lack IgM and IgG antibodies to PB19. Nearly all patients with PB19-PRCA respond to treatment with intravenous Ig therapy with a concomitant decrease in viral titers and a rise in the hemoglobin to levels appropriate for the clinical condition of the patient [17]. Patients may develop fifth disease symptoms after antibody treatment [18].

Immunocompromised Patients

Immunocompromised Patients are those patients who are receiving chemotherapy in case of cancer or immunosuppressive drugs especially in organ transplantation as in liver, kidney or BM transplantation.

PB19 Infection in Transplantation

The first report of PB19 infection after transplantation was published in 1986 [19]. Since then, numerous cases of PB19 infections after solid-organ transplan-tation (SOT) and hematopoietic stem cell transplantation (HSCT) have been reported. Anemia is the predominant clinical manifestation. However, PB19 has also been associated with hepatitis, pneumonitis, myocarditis, and allograft dysfunction. Nonetheless, the full spectrum of clinical manifestations of PB19 infection among transplantat recipients is not well characterized.

The suppression of the RBC population that clinically results in anemia as the hallmark of PB19 infection is consistent with the cellular tropism of this virus [20]. PB19 infects erythroid progenitor cells by binding to the receptor known as the P antigen [20]. Subsequent PB19 replication in erythroid progenitor cells leads to cellular lysis [21], which is characteristically manifested as pure red cell aplasia on bone marrow examination.

It is well known that immunocompetent individuals respond to PB19 infection by producing virus-specific immunoglobulins [22, 23, 24]. Experimental studies have demonstrated that the generation of PB19- specific immunoglobulins (Ig) is temporally accompanied by reduction in the degree of parvoviremia [24]. The impairment in immunity that results from pharmacologic immunosuppression limits the ability of transplant patients to produce neutralizing antibody, which leads to persistent PVB19 infection that manifests as chronic anemia.

Not surprisingly, almost all patients in transplants patient's series had chronic anemia, many patients did not possess PB19-specific Ig at the onset of clinical disease, and almost all transplant patients without PB19 IgM had parvoviremia. Demonstrates that the spectrum of clinical illness related to PB19 is broad. This reflects the ability of PB19 to infect other cells [25]. The cardiotropism of PB19 is suggested by its association with myocarditis and dilative cardiomyopathy or idiopathic left ventricular dysfunction (26, 27, 28, 29), and peripartum cardiomyopathy (3) and by the demonstration of PB19 DNA in fetal myocardial cells [30]. These data support the suggestion that myocarditis may occur in transplant patients with PB19 disease, and this may be misdiagnosed as acute rejection and could result in death from cardiogenic shock [31,32,33]. The most likely cardiac target of PB19 is the endothelium[30,34] because endothelial cells in small cardiac vessels also carry P antigen [35]. Likewise, endothelial infection could serve as the mechanism for PB19-associated thrombotic microangiopathy [35]. Studies of parvoviruses that infect animals demonstrate the virions in various organs [36]. Parvovirus related to Aleutian mink disease was detected in alveolar cells in mink with acute interstitial pneumonitis [37]. Intact Aleutian mink disease parvoviral DNA has also been detected in glomeruli [37]. These animal data support the suggestion that PB19 is a potential cause of pneumonitis [38-40], hepatitis [41-45], and collapsing glomerulopathy [46] in humans. Indeed, PB19 has been demonstrated in the renal tissue and blood of patients with collapsing glomerulopathy and in hepatocytes of a patient with fibrosing cholestatic hepatitis [42]. Nevertheless, the reported associations between PB19 and organ-specific syndromes do not definitely indicate causality.

The inability of transplant patients to mount sufficient anti-PB19 Ig could present a diagnostic dilemma and delay treatment in patients seen at centers who rely on serological examination for the diagnosis PB19. Most of patients already have non detectable PB19 IgM but have positive PCR assay results, suggesting the clinical utility of this molecular assay. So, it is obvious that a negative PB19 IgM serological test result does not rule out the diagnosis of PB19 infection, and PCR should be used whenever a diagnosis of acute PB19 infection is suspected in immunocompromised patients. Among patients who are highly suspected to have PB19 disease but whose peripheral blood PCR assay result is negative, the diagnosis may be confirmed by bone marrow examination.

If feasible, reduction in immunosuppression regimen should be a part of the treatment of PB19 disease. Theoretically, this would allow the immune system to mount specific immunity against PB19. The observation that parvoviremia ceases with generation of Ig [46] led to the current practice of intravenous Ig treatment of PB19. Intravenous Ig contains PB19-specific antibodies. However, the dose and duration of treatment are not standardized. Clinical relapses are commonly observed (i.e. 1 relapse occurs for every 4–5 patients treated), which suggests that the patient experiences a continuous state of severe immunosuppression and there is a need to further reduce immunosuppression drugs or administer intravenous Ig for a longer period to neutralize parvoviremia. The rarity of this infection limits the conduct of prospective trial to assess the optimal dose and duration of treatment.

In conclusion, PB19 can cause significant infectious complication after transplantation. The predominant clinical manifestation of PB19 disease is chronic anemia, although organ invasive manifestations, such as hepatitis, myocarditis, and pneumonitis, can be observed. However, whether these organ specific syndromes are causally linked to PB19 infection remains to be proven. A high index of suspicion is advised when patients present with refractory and severe anemia after transplantation. In this clinical setting, PB19 infection is to be thought in the differential diagnosis, together with the other, more common causes, such as an adverse reaction to treatment, blood loss, and anti-erythropoietin antibody, among others. In this regard, PCR may be a more useful noninvasive test for the confirmation of the diagnosis, because the PB19 serological test results of many transplant patients are negative at the onset of clinical disease.

A retrospective study of parvovirus B19 antibody titres 2 to 3 years after bone marrow transplantation showed persisting IgG, suggesting that persistence of PB19 antibody depends on prior recipient, but not donor, immunity [47].

Allogeneic peripheral blood stem cell transplantation (PBSCT) consistently results in severe immunodeficiency. It is known that human parvovirus B19 can persist in red blood cell precursors in the bone marrow of immunocompromised patients. An infection can cause severe complications including chronic bone marrow failure or pure red cell aplasia because of the inability of patients to produce neutralizing antibodies against the virus[48]. Also, Parvovirus B19 is a cause of delayed red blood cell (RBC) engraftment after marrow transplantation (BMT). The diagnosis of parvovirus infection requires serologic and DNA testing in the context of clinical disease and characteristic marrow morphologic findings; however, the source of infection is often difficult to determine [49].

The major route of transmission of parvovirus B19 is inhalation of respiratory droplets from infected people. Both of these patients were treated in special air-filter rooms designed to create a pathogen-reduced environment and no clinical infections were in family members or hospital staff. There is evidence for parvovirus B19 transmission via blood products, autologus PBSC or bone marrow [50,51]. Patients required erythrocyte infusions during conditioning before transplantation. Parvovirus B19 is known to be a frequent contaminant of blood products[52] but as parvovirus B19 screening is not part of the routine control of blood products is not routinely followed in many countries and so infection via this route cannot be excluded. Another route of infections in those patients is reactivation of old infection. with parvovirus B19 during immunosuppressive conditioning and further immunosuppression. This concept is supported by several reports both in PBSCT and in solid organ transplantation[53].

Parvovirus B19-induces erythematous infection in immuncompromised patients early after PBSC transplantation and should be considered in the differential diagnosis of acute GvHD of the skin. It highlights the importance of excluding the possibility of a viral infection before initiating treatment for acute GvHD. Furthermore, parvovirus B19 infection should be considered in cases of late anaemia after PBSC or bone marrow transplantation, occurring in patients known to be seropositive for parvovirus B19 IgG.

PB19 can cause acute or chronic aplastic anemia in solid-organ transplant recipients and can contribute to some cases of rejection [54]. The overall prevalence of PB19 infection in these patients is about 1-2% during the first year after transplantation. The most common symptom is anemia, but leucopenia and thrombocytopenia have also been observed. Rare cases of hepatic dysfunction, myocarditis, vasculitis and respiratory failure have also been reported. The onset of anemia after transplantation varies from 2 to 34 months [55].

There are a number of reasons why patients who have undergone a solid organ or BM transplant are at risk of developing a PB19 infection:

- There are different paths of infection including airway transmission, blood transfusion, viral reactivation or infection through the transplanted organ. Therefore infection may be acquired from the transplanted organ or blood transfusions or it may represent activation of latent infection [54].
- Immunosuppressive medications are often used with transplant patients and are a major factor in allowing chronic PB19 infection to become established in organ transplant patients [56].
- The cellular receptor for PB19, the P antigen, can be found on myocardial cells which might explain why in some patients PB19 directly infects and damages the heart [57].
- There are reported cases where diseases such as pneumonia or liver disease have been caused by PB19 infection following cardiac transplantation [58].

PB19 DNA was found in 0.8% of patients with myocarditis & 3% of the transplant rejection patients. This is a statistically significant number of transplant rejection patients with evidence of PB19 DNA in comparison to the control group. Two of these transplant rejection patients suffered persistent rejection despite aggressive anti-rejection therapy. These findings show that PB19 infection following transplantation may contribute to some cases of rejection. It also appears that parvoviral myocarditis, although rare, may be more common than previously identified and reported [58].

It is important that PB19 infection is considered in cases where patients present with myocarditis or transplant rejection due to unknown etiological agents. In solid-organ transplant recipients, a PB19 infection should be suspected in the presence of aplastic anemia following specific symptoms of viral infection (fever, fatigue, coetaneous rash), or even if no such symptoms are detected, so it is recommended that a search for PB19 DNA is included in the first-line investigations in any transplanted patients with unexplained anemia. Intravenous Ig therapy is effective in treating chronic PB19 infection in transplant patients [54].

Renal transplant recipients can develop symptomatic PB19 infections as a result of primary infection acquired via the usual respiratory route or via the transplanted organ, or because of reactivation of latent or persistent viral infection. The most common manifestations of PB19 infection in immunosuppressed patients are pure red cell aplasia and other cytopenias. Thus, this diagnosis should be considered in transplant recipients with unexplained anemia and reticulocytopenia or pancytopenia. Collapsing glomerulopathy and thrombotic microangiopathy have been reported in association with PB19 infection in renal transplant recipients, but a causal relationship has not been definitively established. Prompt diagnosis of PB19 infection in the renal transplant recipient requires a high index of suspicion and careful selection of diagnostic tests, which include serologies and polymerase chain reaction. Most patients benefit from intravenous immunoglobulin therapy and/or alteration or reduction of immunosuppressive therapy. Conservative therapy might be sufficient in some cases [59].

Further more, case reports implicate parvovirus in the pathogenesis of proliferative glomerulonephritis and collapsing glomerulopathy, but a causal relationship has not been established. A proposed role for B19 infection is based on the temporal association of renal findings with viral infection, positive serology, and identification of the viral genome in the glomerulus. Mechanisms may include cytopathic effects on glomerular epithelial cells and/or endothelial cells and glomerular deposition of immune complexes. Patients who require dialysis may have increased susceptibility to acute and chronic anemia after parvoviral infection. Factors that predispose this population to complications of PB19 infection include impaired immune response, deficient erythropoietin production, and possibly decreased erythrocyte survival. The clinical burden of parvovirus B19 infection in renal transplant recipients may be underestimated; these individuals may develop persistent viremia as a result of a dysfunctional immune response. Chronic anemia and pure red blood cell aplasia are the most common complications of parvovirus infection in this population; the diagnosis should be considered in transplant recipients with unexplained anemia or pancytopenia. Allograft rejection and dysfunction have been reported in association with infection, but a cause-effect relationship has not been proved. Further investigation of the relationship between B19 and kidney disease is warranted [60].

PB19 Infection in Patients Receiving Chemotherapy

Immunocompromised children, including those undergoing chemotherapy treatments of malignant disease; are at particular risk for infection with PB19. However, these patients' attenuated immune responses may obscure the serologic and clinical manifestations of the infection [61]. Chronic infection with of PB19 tends to occur in these patients and manifests as PRCA, chronic anemia and to less extent by thrombocytopenia, neutropenia and pancytopenia [62].

A research was done to present descriptive serologic study or clinical and hematological implications of PB19 in children with acute lymphoplastic leukemia (ALL) from time of initial admission until discontinuation of chemotherapy. 75 patients were studied by PCR and ELISA, researchers found that 8% of PB19-seronegative patients seroconverted and infection triggered profound anemia and thrombocytopenia. PB19-specific IgG disappeared in 26% of PB19-seropositive patients and these patients were significantly younger and the PB19 IgG titers were lower on admission compared with patients who continuously displayed PB19 IgG. PB19 DNA was detected in the seroconverted patients and this helped in determining the time of infection, which coincided with a PB19 epidemic in 75% of patients. Typical presentation of PB19 infection (rash & arthralgia) was observed in only one patient and PB19 infection was able to mimic a leukemic relapse or therapy induced cytopenia in other patients Heegaard et al. (1999) reported a three-years old boy diagnosed with ALL. Two years later, while on maintenance chemotherapy, the patient was readmitted because of thrombocytopenia. The thrombocytopenia was proved to be a sequel of PB19 infection as evidenced by the finding of specific DNA in serum and BM samples.

In these cases multiple transfusions of red blood cells or platelets and cessation of maintenance chemotherapy for up to 3 weeks may be beneficial [63]. This high infection rate may be due to susceptibility of immunocompromised patients to infection. Data suggests that cancer patients younger than 40 years who receive multiple courses of systemic chemotherapy have an increased risk of PB19 infection [65].

Conclusion

Parvovirus B19 infection can be asymptomatic or can cause a broad range of diseases, including diseases found among normal, nonimmunocompromised hosts (erythema infectiosum, arthropathy, hydrops fetalis); hematologic diseases in immunocompromised hosts (aplastic crisis, chronic anemia, idiopathic thrombocytopenic purpura, transient erythroblastopenia of childhood, Blackfan-Diamond anemia). Moreover, it can be associated with serious complication in pateints with transplantation. Less common clinical associations of PB19 virus infection include various skin eruptions, hematologic disorders such as neutropenia, and renal affection.

References

[1] Mustafa MM and McClain KL. Diverse haematologic effects of parvovirus B19 infection. *Pediat. Haematol.* 1996, 43:809-21.

[2] Young NS and Brown KE. Parvovirus B19. *N. EJM.* 2004,350:586-97.

[3] Kurtzman GJ, Ozawa K, Cohen B et al. Chronic bone marrow failure due to persistent B19 parvovirus infection. *N. EJM.* 1987; 317: 287±294.

[4] Frickhofen N, Abkowitz JL, Safford M et al. Persistent B19 parvovirus infection in patients infected with human immunodeficiency virus type-1 (HIV-1): a treatable cause of anemia in AIDS. *An. of Intern. Med.* 1990; 113: 926±933.

[5] Kurtzman GJ, Cohen B, Meyers P et al. Persistent B19 parvovirus infection as a cause of severe chronic anaemia in children with acute lymphocytic leukaemia. *Lancet.* 1988; ii: 1159±1162.

[6] Kurtzman G, Frickhofen N, Kimball J et al. Pure red-cell aplasia of 10 years' duration due to persistent parvovirus B19 infection and its cure with immunoglobulin therapy. *NEJM.* 1989; 321: 519±523.

[7] Pont J, Puchhammer-StoÈ ckl E, Chott A et al. Recurrent granulocytic aplasia as clinical presentation of a persistent parvovirus B19 infection. *Br. J. of Haematol.* 1992; 80: 160±165.

[8] Brown K. Haematological consequences of parvovirus B19 infection. *Clin. Haematol.* 2000, 13: 2/ 245-259.

[9] Abkowitz JL, Brown KE, Wood R W et al. Human parvovirus B19 (HPV B19) is not a rare cause of anemia in HIV-seropositive individuals. *Clinical Research.* 1993; 41: 393A.

[10] Tavil B, Sanal O, Turul T, Yel L, Gurgey A, Gumruk F. Parvovirus B19-induced persistent pure red cell aplasia in a child with T-cell immunodeficiency. *Pediatr. Hematol. Oncol.* 2009 ;26:2/63-8.

[11] Davidson JE, Gibson B, Gibson A, et al. Parvovirus infection, leukaemia and immunodeficiency. *Lancet.* 1989. I:102-105.

[12] Gahr M, Pekrun A and Eiffert H. Persistence of parvovirus B19 DNA in blood of a child with severe combined immunodeficiency associated with chronic pure red cell aplasia. *Eur. J. Pediatr.* 1991, 150: 470-72.

[13] Brown KE and Young NS. Parvoviruses & Bone Marrow Failure. *Stem. cells.* 1996, 14:151-63.

[14] Setubal S, Jorge_Pereira MC, de Sant'anna AL, et al. Clinical presentation of parvovirus B19 infection in HIV-infected patients with & without AIDS. *Rev. Soc. Bras. Med. Trop.* 2003, 36:299-302.

[15] De Renzo A, Azzi A, Zakrzewska K, Cicoira L, Notaro R, Rotoli B. Cytopenia caused by parvovirus in an adult ALL patient. *Hematologica.* 1994; 79:259–261.

[16] Clarke J, Lee JD. Primary human parvovirus B19 infection in an HIV infected patient on highly active antiretroviral therapy. *Sex Transm. Infect.* 2003; 79:336

[17] Koduri PR. Parvovirus B19-related anemia in HIV- infected patients. *AIDS Patients Care. STDS.* 2000,14:7-11.

[18] Thio K, Janner D. Aplastic anemia in a cardiac transplant recipient. *Pediatr. Infect. Dis. J.* 1996; 15:1139:1141–2.

[19] Neild G, Anderson M, Hawes S, Colvin BT. Parvovirus infection after renal transplant. *Lancet.* 1986; 2:1226–7.

[20] Brown KE, Anderson SM, Young NS. Erythrocyte P antigen: cellular receptor for B19 parvovirus. *Science.* 1993; 262:114–7.

[21] Young N, Harrison M, Moore J, Mortimer P, Humphries RK. Direct demonstration of the human parvovirus in erythroid progenitor cells infected in vitro. *J. Clin. Invest.* 1984; 74:2024–32.

[22] Poblotzki AV, Gerdes C, Reischl U, Wolf H, Modrow S. Lymphoproliferative responses after infection with human parvovirus B19. *J. Virol.* 1996, 70:10, 7327–7330.

[23] Brown K. E., Kurtzman GJ, Cohen BJ, Field AM et al. Immune response to B19 parvovirus and an antibody defect in persistent viral infection. *J. Clin. Invest.* 1989; 84: 1114-1123.

[24] Mortimer PP, Humphries RK, Moore JG, Purcell RH, Young NS. A human parvovirus-like virus inhibits haematopoietic colony formation in vitro. *Nature.* 1983; 302:426–9.

[25] Soderlund-Venermo M, Hokynar K, Nieminen J, Rautakorpi H, Hedman K. Persistence of human parvovirus B19 in human tissues. *Pathol. Biol.* 2002; 50:307–16.

[26] Porter HJ, Quantrill AM, Fleming KA. B19 parvovirus infection of myocardial cells. *Lancet.* 1988; 1:535–6.

[27] Naides SJ, Weiner CP. Antenatal diagnosis and palliative treatment of non-immune hydrops fetalis secondary to fetal parvovirus B19 infection. *Prenat. Diagn.* 1989; 9:105–14.

[28] So B. J. Chae K M., Lee K K., Lee Y J. Jeong B H. Pure red cell aplasia due to parvovirus B19 infection in a renal transplant patient: a case report. *Transplantation Proceedings.* 2000, 32: 7/ 1954-1956

[29] Pankuweit S, Ruppert V, Eckhardt H, Strache D, Maisch B. Pathophysiology and aetiological diagnosis of inflammatory myocardial diseases with a special focus on parvovirus B19. *J. Vet. Med. B Infect. Dis. Vet. Public Health.* 2005; 52:344–7.

[30] Bultmann BD, Klingel K, Sotlar K, et al. Fatal parvovirus B19-associated myocarditis clinically mimicking ischemic heart disease: an endothelial cell-mediated disease. *Hum. Pathol.* 2003; 34:92–5.

[31] Dettmeyer, R., R. Kandolf, A. Baasner, S. Banaschak, A. M. Eis-Hubinger, and B. Madea. Fatal parvovirus B19 myocarditis in an 8-year-old boy. *J. Forensic. Sci.* 2003, 48:183–186.

[32] Schowengerdt KO; Ni J; Denfield SW; Gajarski RJ; Bowles NE; Rosenthal G; Kearney DL; Price JK; Rogers BB; Schauer GM; Chinnock RE; Towbin JA. Association of parvovirus B19 genome in children with myocarditis and cardiac allograft rejection: diagnosis using the polymerase chain reaction. *Circulation.* 1997, 18; 96:10/3549-54.

[33] Rosenthal D L, Kearney J K, Price B B, Rogers G M, Schauer R, Chinnock E, Towbin J A. Association of parvovirus B19 genome in children with myocarditis and cardiac allograft rejection: diagnosis using the polymerase chain reaction. *Circulation.* 1997, 96:3549–3554.

[34] Bultmann BD, Sotlar K and Klingel K. Parvovirus B19. *NEJM.* 2004; 350:2006–7.

[35] Murer L, Zacchello G, Bianchi D, et al. Thrombotic microangiopathy associated with parvovirus B 19 infection after renal transplantation. *J. Am. Soc. Nephrol.* 2000; 11:1132–7.

[36] Alexandersen S, Bloom ME. Studies on the sequential development of acute interstitial pneumonia caused by Aleutian disease virus in mink kits. *J. Virol.* 1987; 61:81–6.

[37] Alexandersen S, Bloom ME, Wolfinbarger J, Race RE. In situ molecular hybridization for detection of Aleutian mink disease parvovirus DNA by using strand-specific probes: identification of target cells for viral replication in cell cultures and in mink kits with virus-induced interstitial pneumonia. *J. Virol.* 1987; 61:8/ 2407–2419

[38] Klumpen HJ, Petersen EJ, Verdonck LF. Severe multiorgan failure after parvovirus B19 infection in an allogeneic stem cell transplant recipient. *Bone Marrow Transplant.* 2004; 34:469–70.

[39] Janner D, Bork J, Baum M, Chinnock R. Severe pneumonia after heart transplantation as a result of human parvovirus B19. *J. Heart Lung Transplant.* 1994; 13:336–8.

[40] Rosenthal G, Kearney D L, Price J K, Rogers B B, Schauer G M, Chinnock R E, Towbin J A. Association of parvovirus B19 genome in children with myocarditis and cardiac allograft rejection: diagnosis using the polymerase chain reaction. *Circulation.* 1997, 96:3549–3554

[41] Moreux N, Ranchin B, Calvet A, Bellon G, Levrey-Hadden H. Chronic parvovirus B19 infection in a pediatric lung transplanted patient. *Transplantation.* 2002; 73: 565–8.

[42] Schleuning M, Jager G, Holler E, et al. Human parvovirus B19-associated disease in bone marrow transplantation. *Infection.* 1999; 27: 114–7.

[43] Shan YS, Lee PC, Wang JR, Tsai HP, Sung CM, Jin YT. Fibrosing cholestatic hepatitis possibly related to persistent parvovirus B19 infection in a renal transplant recipient. *Nephrol. Dial. Transplant.* 2001; 16: 2420–2.

[44] Lee PC, Hung CJ, Lei HY, Chang TT, Wang JR, Jan MS. Parvovirus B19-related acute hepatitis in an immunosuppressed kidney transplant. *Nephrol. Dial. Transplant.* 2000; 15:1486–8.

[45] Keung YK, Chuahirun T, Wesson D. Concomitant parvovirus B19 and cytomegalovirus infections after living-related renal transplantation. *Nephrol. Dial. Transplant.* 1999; 14:469–71.

[46] Moudgil A, Shidban H, Nast CC, et al. Parvovirus B19 infection-related complications in renal transplant recipients: treatment with intravenous immunoglobulin. *Transplantation.* 1997; 64: 1847–50.

[47] Mortimer PP, Humphries RK, Moore JG, Purcell RH, Young NS. Ahuman parvovirus-like virus inhibits haematopoietic colony formation in vitro. *Nature.* 1983; 302: 426–9.

[48] Ang HA, Apperley JF, Ward KN. Persistence of antibody to human parvovirus B19 after allogeneic bone marrow transplantation: role of prior recipient immunity. *Blood.* 1997; 89: 4646–4651.

[49] Solano C, Juan O, Gimeno C, Garcia-Conde J. Engraftment failure associated with peripheral blood stem cell transplantation after B19 parvovirus infection. *Blood.* 1996; 88: 1515–1517.

[50] Arnold DM, Neame PB, Meyer RM, Soamboonsrup P, Luinstra KE, O'Hoski P, Garner J, Foley R. Autologous peripheral blood progenitor cells are a potential source of parvovirus B19 infection. *Transfusion,* 2005, 45:3/394-8

[51] Heegaard ED, Petersen BL. Parvovirus B19 transmitted by bone marrow. *Br. J. Haematol.* 2000; 111: 659–661.

[52] Plentz A, Hahn J, Knöll A, Holler E, Jilg W, Modrow S. Exposure of hematologic patients to parvovirus B19 as a contaminant of blood cell preparations and blood products. *Transfusion.* 2005; 45: 1811–1815.

[53] Bonvicini F, Gallinella G, Gentilomi GA, Ambretti S, Musiani M, Zerbini M. Prevention of iatrogenic transmission of B19 infection: different approaches to detect, remove or inactivate virus contamination. *Clin. Lab.* 2006; 52: 263–268.

[54] Eid AJ, Brown RA, Patel R, Razonable RR. Parvovirus B19 infection after transplantation: a review of 98 cases. *Clin. Infect. Dis.* 2006; 43: 40–48.

[55] Murer L, Zacchello G, Bianchi D, et al. Thrombotic Microangiopathy associated with Parvovirus B19 Infection after Renal Transplantation. *J. Am. Soc. Nephrol.* 2000,11:1132-37.

[56] Broliden K. Parvovirus B19 Infection in pediatric solid organ & bone marrow transplantation. *Pediatr. Transplant.* 2001,5: 320-30.

[57] Moudgil A, Shidban H, Nast CC, et al. Parvovirus B19 Infection-Related Complications in Renal Transplant Recipients. *Transplantation.* 1997,64:1847-50.

[58] Heegaard ED, Mayer J, Hornsleth A, et al. Parvovirus B in patients with chronic anemia. *Haematologica.* 1997, 82: 402-5.

[59] Schowengerdt KO; Ni J; Denfield SW; Gajarski RJ; Bowles NE; Rosenthal G; Kearney DL; Price JK; Rogers BB; Schauer GM; Chinnock RE; Towbin JA . Association of parvovirus B19 genome in children with myocarditis and cardiac allograft rejection: diagnosis using the polymerase chain reaction. *Circulation.* 1997, 18;96:10/3549-54.

[60] Waldman M, Kopp JB.Parvovirus-B19-associated complications in renal transplant recipients. *Nat. Clin. Pract. Nephrol.* 2007; 3:10/540-50.

[61] Mc Nall RY, Head DR, Pui CH, et al. Parvovirus B19 infection in a child with acute lymphoblastic leukaemia durig induction therapy. *J. Pediatr. Hematol. Oncol.* 2001, 23:309-11.

[62] El-Mahallawy HA, Mansour T, El-Din SE, et al. Parvovirus B19 infection as a cause of anaemia in pediatric acute lymphoblastic

leukaemia patients during maintenance chemotherapy. *J. Pediatr. Hematol. Oncol.* 2004,26:403-6.

[63] Heegaard ED and Schmiegelow K. Serologic study on parvovirus B19 infection in childhood acute lymphoblastic leukaemia patients during chemotherapy: clinical & hematologic implications. *J. Pediatr. Hematol. Oncol.* 2002, 24: 368-73.

[64] Heegaard ED, Kerndrup GB, Carlsen NT et al. Thrombocytopenia caused by Parvovirus B19 infection in a child with acute lymphatic leukemia. *Ugeskr Laeger.* 1999, 16:6501-2.

[65] Fattet S, Cassinotti P and Popovic MB. Persistent Human Parvovirus B19 Infection in Children Under Maintenance Chemotherapy for Acute Lymphocytic Leukemia. *J. of Ped. Hemat. Oncol.* 2004, 26: 497-503.

Chapter 4

HEMATOLOGICAL CONSEQUENCES OF PB19 INFECTION

ABSTRACT

Parvovirus B19, a member of the Erythrovirus genus, is the only member of the Parvoviridae family known to be pathogenic in humans. Erythroviruses are so named because of their tropism and selective replication in erythroid progenitor cells. Haematological consequences of B19 infection arise due to a direct cytotoxic effect on erythroid progenitors in bone marrow with interruption of erythrocyte production. This result in what is called transient aplastic attack. Other important issue about PB 19 is its association acute leukemia with hematological consequences before development of the disease, during and after chemotherapy.

Keywords: PB19 pathogenesis, hematological consequences.

INTRODUCTION

Certain populations are vulnerable to complications from the interrupted erythropoiesis that arises with infection.

These populations include people with hematologic disorders that cause an increased rate of erythrocyte destruction, such as hereditary spherocytosis, thalassaemias, sickle cell anemia's, glucose 6 phosphate deficiency, chronic autoimmune hemolytic anemia, cold and heat antibody-mediated autoimmune hemolytic disease, paroxysmal nocturnal hemoglobinuria, iron deficiency anemia and even blood loss [1].

In those patients with increased erythropoiesis, the transient cessation of red cell production which is asymptomatic in normal individuals can become significant, and precipitate an `aplastic crisis'. The term aplastic crisis was coined by Owren [2] to describe the abrupt onset of severe anaemia with absent reticulocytes in patients with hereditary spherocytosis, in contrast to haemolytic crises in which bone marrow turnover is increased and reticulocytes are high, but may occur in any setting where there is increased red cell production.

Transient aplastic crises (TAC) can occur under conditions of erythroid `stress': such as haemorrhage, iron deficiency anaemia and following kidney or bone marrow transplantation. Aplastic crisis can be the first presentation of an underlying haemolytic disease in a hematological well compensated patient, and although suffering from an ultimately self-limiting disease, patients with aplastic crisis can be severely ill. Symptoms may include dyspnoea, lassitude and even confusion due to the worsening anemia. Congestive heart failure and bone marrow necrosis may develop, and the illness can be fatal.

TAC due to PB19 infection has now been described in a wide range of patients with underlying haemolytic disorders, including hereditary spherocytosis, thalassaemia, red cell enzymopathies such as pyruvate kinase deficiency, and autoimmune haemolytic anaemia [3].

Community acquired aplastic crisis is almost always due to parvovirus B19 and parvovirus infection should be the presumptive diagnosis in any patient with anemia due to abrupt cessation of erythropoiesis as documented by reduced reticulocytes and the characteristic bone marrow morphology. In contrast to patients with erythema infectiosum, PB19 infection in these patients results in TAC which is characterized by reticulocytopenia, absent erythroid precursors in the bone marrow (BM) and transient worsening of anemia [4].

Although the erythrocytes are predominantly affected with presentation often of a pure red cell aplasia; concurrent thrombocytopenia, neutropenia, or pancytopenia is found infrequently especially in patients with functioning spleen [5].

Hematological consequences of PB19 infection are mainly due to a direct cytotoxic effect on erythroid progenitors in bone marrow (BM) with interruption of erythroid production, figure 1 [6].

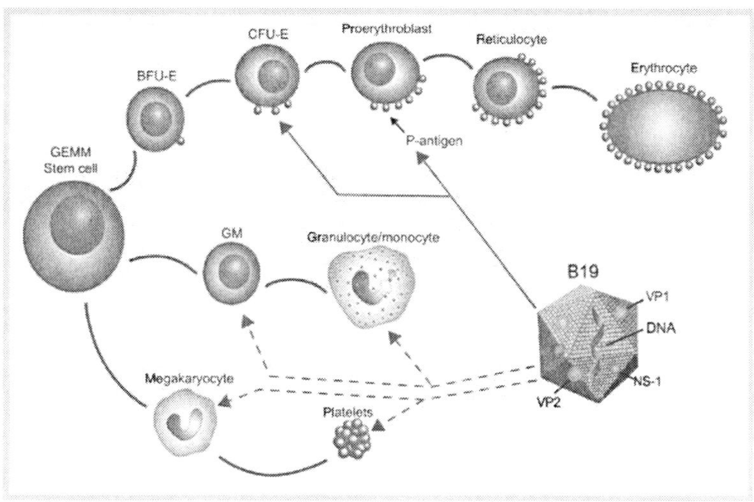

Figure (5) PB19 binds to immature erythroblasts thereby arresting production of mature erythropoietic cells

TAC patients are often viraemic at the time of presentation with concentrations of virus as high as 10^{14} genome copies/ml. The diagnosis therefore is readily made by detection of PB19 DNA in the serum. As the PB19 DNA is cleared from the serum, PB19 specific IgM becomes detectable. TAC is a unique event in the patient's life, and following the acute infection immunity is life-long.

Pathogenesis of PB19 Related to Hematological Disorders

The infection cascade of PB19 involve binding of the virus to cell receptors. The cellular receptor for PB19 infection has been regarded as blood group P antigen based on the failure of PB19 infection in a patient with hereditary defect of P antigen [7]. The notion that PB19 receptor is not solely P antigen may be compatible with clinical finding that PB19 has been detected in mononuclear cells of blood or tonsils at acute and/or prolonged PB19 infection [8, 9]. Also, following PB19 infection, the numbers of peripheral blood lymphocytes may decrease despite undetectable levels of P antigen on their cell surface [10, 11]. Finally, autoimmune-like phenomena including anti-nuclear antibodies, rheumatoid factors or anti-phospholipid antibodies are often associated with B19 infection [9, 12], and the levels of tumour necrosis factor (TNFα) and interferone gamma (IFNγ) secreted from macrophages or T cells are elevated during acute and/or prolonged PB19 infection [13].

So it is obvious that the target cells of PB19 may be not be exclusively P antigen-positive erythroid-lineage cells, as illustrated by the poor relationship between P antigen expression levels and the efficiency of PB19 infection [14] or the failure of PB19-binding to globoside [15]. Another presumed receptor is integrin. The supposed is α5β1 integrin for PB19 [16].

In addition recently discovered receptor for PB19 is Ku receptor. Ku is a heterodimeric DNA-binding protein consisting of a 70 kDa (Ku70) and an 80 kDa (Ku80) subunit, and was originally identified as a nuclear antigen recognized by autoantibodies in patients with systemic lupus erythematosus and scleroderma [17]. Ku has a central role in multiple nuclear processes, including DNA repair, chromosome maintenance, transcription regulation and V (D) J recombination. Ku is abundant in the nucleus, consistent with its function as a DNA-protein kinase (DNA-PK) [18, 19]. Also, recent studies have shown cytoplasm or surface localization of Ku in various types of cells, including of leukemia, multiple myeloma, and tumor cell lines. Ku is a component of the DNA-PK complex in membrane rafts of mammalian cells [18].

Although the role of surface Ku80 has not been well clarified [20], signal transduction and Ku80 are coupled in both B and T lymphocyte cells [17, 20], and localization of the DNA-PK complex in lipid raft suggests a putative role in the signal transduction pathway following ionizing radiation [18].

It is recently reported that Ku interacts with metalloproeinase 9 at the cell surface of highly invasive hematopoietic cells of normal and tumor cell origin, and Ku80/MMP-9 interaction at the cell membrane may result in contribution to the invasion of tumor cells through regulation of extracellular matrix remodeling [21]. Further function attributed to the membrane form of Ku, which is expressed by hypoxia, that it mediates cell adhesion of plasma cells [21, 22], indicating a role for Ku as an adhesion receptor for fibronectin [23].

Data implicate Ku80 as a coreceptor involved in PB19 infection. This suggestion may explain clinical findings associated with PB19 infection to non-erythroid cells as Ku80 is also expressed on the surface of immune cells in bone marrow in vivo.

Therefore, PB19 infection often associated with decrease in number of leukocytes or lymphocytes in blood during acute PB19 infection, as well as increased levels of TNFα and IFNγ in blood.

PB19 may affect immune cells in bone marrow or the synovium. Its persistence lead to secrete an inflammatory cytokine through the activation of antigen presenting cell type 1 (AP1) and antigen presenting cell type 2 (AP2) by PB19 non specific protein 1 (NS1) [24]. Stimulation of cellular receptors with PB19 may trigger activation of signal cascades in host cells, which may explain why immune cells in acute and prolonged PB19 infection are functionally altered.

After PB19 entrance into erythroblasts death of cell occur due to apoptosis.

Erythroblasts apoptosis due to PB19 infection of erythroid lineage cells is characterized by a gradual cytocidal effect mediated by NS1 protein, [25] and by features of apoptosis [26]. It has recently been shown that NS1 is responsible for apoptosis of erythroid cells. This is approved by abolishing the apoptosis by nucleotide mutation within the nucleoside triphosphate binding site of NS1 [27].

Non specific protein 1, a compenent of PB19 viral genome, mediates apoptosis by a pathway that involves caspases 3, 6, and 8 and seems to be mediated by an increase in sensitivity to apoptosis induced by TNFα. In addition, B19 infection induces cell cycle arrests at the G (1) and G (2) phases. The G (1) arrest is induced by NS1 expression prior to apoptosis induction in PB19-infected cells, while the G (2) arrest is induced not only by infectious PB19 but also by UV-inactivated PB19, which lacks the ability to express NS1 [28].

Figure 4 summarizes affection of erythropoietic cells by PB19 infection.

PB19 binds to immature erythroblasts thereby arresting production of mature erythropoietic cells. Following acute infection, the reticulocyte count in peripheral blood is zero and if the patients have an underlying disorder with pathologic red cell survival, the number of erythocytes may fall dramatically in peripheral blood. The pathogenesis of thrombocytopenia is thought to be explained by the cytotoxicity of the NS1 protein [29].

Transient Aplastic Crisis (TAC)

Aplastic anaemia is a BM haemopoietic failure induced by a variety of causes such as certain chemicals, drugs, radiation and virus infections. Association of viral infection with aplastic anemia is well documented. PB19 has been shown to cause TAC [30]. TAC due to PB19 has been described in a wide range of patients with underlying haemolytic disorders including red cell membrane defects as hereditary spherocytosis & hereditary stomatocytosis, haemoglobinopathies as thalassaemias & sickle cell anaemias, red cell enzymopathies as pyruvate kinase deficiency & glucose 6 phosphate deficiency, chronic autoimmune hemolytic anemia, cold and heat antibody-mediated autoimmune hemolytic disease, paroxysmal nocturnal hemoglobinuria, iron deficiency anaemia and even blood loss. So PB19 infection should be suspected in any case of anaemia complicated by aplastic crisis [1].

PB19 infection is a serious cause of TAC in patients with hemolytic anemia. Data was found by a research done in USA on patients with sickle cell disease, it showed that 30% had evidence of old PB19 infection at first testing [31].

Higher rates of recent infection were found by other studies, Rao et al., (1992) found approximately 70% of cases of TAC patients with sickle cell disease were caused by acute PB19 infection. Also, Brown and Young (1996) showed that 86% of TAC was associated with recent PB19 infection

Another research done in China proved that PB19 infection is not only correlated with the occurrence of children acute aplastic anaemia and chronic aplastic anaemia, but also with adult acute aplastic anaemia and chronic aplastic anaemia, and might be an important viral cause for aplastic anaemia in humans [34].

Evidence that the TAC of hemolytic anemia is due to PB19 infection depends on many factors: the disease almost never recurs in the same individual and no other specific agent could be detected [35].

It is speculated that TAC does not inevitably develop in every individual with hemolytic anemia who is infected with PB19 [35]. Similarly, Zimmerman et al., (2003) proved that subclinical parvovirus infection appears to be common in children with sickle cell anemia although most patients had no documented previous TAC.

Although the erythrocytes are predominantly affected with presentation often of a pure red cell aplasia; concurrent thrombocytopenia, neutropenia, or pancytopenia is found infrequently especially in patients with functioning spleen [37].

PB19 infection was a relatively common finding in patients suffering from chronic anemia irrespective of the cause. This could be explained on the basis that PB19 viremia may be present for several years in some cases resulting in colonization of the virus in BM causing suppression of erythropoiesis that manifests as PRCA and chronic anaemia [38, 39].

TAC was described as the abrupt onset of severe anemia with reticulopenia in patients suffering from chronic hemolysis due to cessation of erythrocyte production in the BM secondary to the tropism and direct cytotoxic effect of the virus for erythroid progenitor cells [40]. It is the first and commonest illness associated with PB19 infection which in persons with a need for increased red cell production can result in a sever but self-limited reticulocytopenic anaemia or TAC. Even in normal persons, infection with PB19 probably causes a transient arrest in the production of red cells in the BM, but with a normal red cell life span of 120 days; this effect has very little clinical significance. Similarly, fetuses thought to be severely affected by PB19-intrauterine infection in the first & second trimester as the half-life of red blood cells is apparently shorter than RBCs at the BM haematopoietic stage [41].

It is proved that PB19 associated damage of erythroid linage cells is due to cytotoxicity mediated by viral protein NS1 which causes apoptotic changes in the infected cells, these changes are due to cell cycle arrest at the GI & GП phases [42]. The GI arrest is induced by expression prior to apoptosis induction in PB19-infected cells, while the GП arrest is induced not only by infectious agent but also by ultraviolet- inactivated PB19 which lacks the ability to express NS1 [43].

Data reported that constitutional signs such as fever are the most frequent associated clinical findings and that a rash is almost never present in the acute phase of the illness, but splenomegaly was present in (35%) of patients suffering from hemolytic anemia and presented with TAC [45]. The usual presentation includes a prodrome of fever and constitutional symptoms suggesting viral illness from 1-17 days before the aplasia.

So, it is obvious that pain, fever and acute splenic sequestration were more frequent events with acute PB19 infection compared with acute events in uninfected patients with TAC. Shortly afterwards, patients have extreme pallor and fatigue [35]. Fever is considered as a consistent sign that occur with patients infected with PB19 [44]. It is more common during acute PB19 infection. The presence of fever may result from a greater viral load observed in acute infection [45]. On the other hand the occurrence of rash on those patients is uncommon in those patients [46]. Absence of rash in TAC is believed to result from antigen excess and prolonged viremia during most of the infection [33] and because the appearance of the rash corresponds with the development of IgG antibodies which occurs after the viremia has cleared [47]. Another hypothesis is that patients in TAC often receive blood transfusion, so the rash may be not well noticed and attributed to a transfusion reaction [46]. Another clinical sign that may be detected in those patients are arthropathy [48].

The usual findings in cases of TAC there are decrease in hemoglobin, but marked anemia may occur in patients with mild hemolysis. Nevertheless when BM erythrocyte production is transiently shut off by viral PB19 or other infections due to uncompensated destruction of erythrocytes there is profound reticulocytopenia leading to a sudden and often life-threatening fall in serum haemoglobin (may reach below 5gm/dl), so severe disease with heart failure and death is possible [47, 49].

Bone marrow aplasia due to parvovirus B19 infections is usually mild and self-limited. In patients with hereditary or acquired hemolytic anemias, PB19 infection can cause a severe and life-threatening anemia due to the shortened half-life of red cells, but here, too, the transient nature of the infection soon remits the symptoms. lthough recovery usually occurs in 1-2 weeks by appearance of specific antibodies with sequential appearance of peripheral blood normoblasts, reticulocytes and a rising haemoglobin, transfusion is required and may be life saving [47]. TAC is usually a unique event in the life of a patient, suggesting the induction of long-lasting protective immunity [37].

The lack of the typical rash pattern in a large proportion of PB19 and the similarity of clinical manifestations to other rash diseases, specially to rubella and its difficulty to be seen in persons with dark skin highlight the difficulty of diagnosing PB19 infection on clinical grounds alone [50].

So, the diagnosis is readily made by detection of B19 DNA in the serum. As PB19 DNA is cleared from the serum, PB19-specific IgM becomes detectable. BM aspirates reveal marked erythroid hypoplasia with characteristic nuclear inclusions in the giant pronormoblasts [37].

Community acquired aplastic crisis is almost always due to parvovirus PB19 and should be the expected diagnosis in any patient with anaemia due to abrupt cessation of erythropoiesis as documented by reduced reticulocytes and bone marrow appearance [33].

It is reported that prevalence of PB19 IgG antibodies among patients with chronic anemia in different hematological disorders ranges from 60%-70%. [51, 52]. Also, PB19 IgM is detected in significantly higher numbers of investigated patients with aplastic anemia in comparison to the controls. The presence pf PB19 DNA is usually detected in those patients. In contrast to patients with erythema infectiosum, TAC patients are often viremic at the time of presentation with concentrations of virus as high as 10^{14} genome copies/ml.

In contrary, lower rates of positive IgM cases was presented by another research done in Saudi Arabia on several types of chronic hemolytic anemia;

acute PB19 infection detected by specific IgM antibodies was the cause of aplastic crisis in (15%) of cases with slightly higher prevalence of IgG antibodies (68%) [53]. Also, lower rates of anti-human PB19 was reported in Rio de Janeiro which indicates prevalence of anti-human PB19 IgG in about (32.1%) of patients with chronic haemolytic anaemia [54].

PB19 infection in immunocompromised patients differs. Chronic infections with parvovirus are more characteristically associated with immunodeficiency states [55].

Persistent PB19 infection producing pure red cell aplasia with or without chronic bone marrow (BM) failure is the commonest feature of PB19 infection in these patients **[37]**. As a result of that; the usual BM finding in acute PB19 infections are marked erythroid hypoplasia and occasionally giant erythroblasts. However, unusual BM manifestations of PB19 infection in immunocompromised patients were reported by Crook and his colleagues (2000), the BM was hypercellular in most of patients and erythroid precursors were increased with abundant intranuclear inclusions. The inclusions were eosinophilic and compressed the chromatin against nuclear membrane [56].

Idiopathic Thrombocytopenic Purpura (ITP)

Thrombocytopenia associated with PB19 infection may occur before or after the appearance of rash in patients with erythema infectiosum. In children, unlike in adults, most cases of ITP are of acute onset and are often preceded by a specific viral infection [1].

PB19-induced thrombocytopenia seems to consist of a central and a peripheral type. Thrombocytopenia of central origin is due to BM suppression, and the possible cytopathologic effect is underlined by the finding that the NS1 protein, produced by PB19, has been found to inhibit the megakaryocytic colony formation. This indicates tissue tropism of PB19 beyond the erythroid progenitor cell and shows that viral proteins may be toxic to cell populations that are non permissive for viral DNA replication[57]. Destructive thrombocytopenia of peripheral origin may result from immunologically mediated antiplatelet antibody production with subsequent excessive platelet clearance in the reticuloendothelial system [1].

One study demonstrated a current or recent PB19 infection in 6 of 47 pediatric ITP patients (13%), and it was suggested that children with ITP and associated PB19 infection are characterized by acute onset of profound thrombocytopenia [58]. Among the PB19-positive children, duration of disease was brief in three children treated with immunoglobulin but chronic in the remaining three patients given high-dose steroids [1].

Neutropenia

Neutropenia, when occurring during the course of primary PB19 infection in healthy individuals; is moderate and transient. Severe neutropenia has been reported during the course of chronic PB19 infection as observed in patients undergoing myelosuppressive or immunosuppressive therapy. In these conditions; neutropenia is a part of BM failure-related pancytopenia [59].While the pathophysiology of PB19-related erythroid cytotoxicity is now well documented; that of neutropenia remains unclear. It has been suggested that neutropenia may be either due to that circulating neutrophils could carry replicating parvovirus in acutely infected patients or may originate from an autoimmune process as shown by the decrease in cloning efficiency observed when granulocytic progenitors derived from cells harvested from BM of children with PB19-induced chronic neutropenia are cultured in vitro in autologus plasma [59].

Primary autoimmune neutropenia occurs predominantly in infancy. The BM of children with neutropenia was examined for PB19 DNA. Results indicated that PB19 infection may be a common cause of immune-mediated neutropenia in childhood (15 of 19 patients) [1].

Heegaard et al., (1997) who investigated out 43 patients suffering from chronic anemia with different hematological diseases such as PRCA, hemolytic anemia without TAC, aplastic anemia, acute myeloid leukemia (AML), myelodysplastic syndrome (MDS) and malignant lymphomas. He revealed that all of them had hemoglobin (Hb) < (9.7 g/dl) with marked reticulocytopenia, but neutrophil and platelet counts revealed no significant differences between positive and negative cases.

Controversy is found about presence of neutropenia in cases of PB19 infection, some authors denied association of PB19 neutropenia on the basis of marked tropism of PB19 to erythroid progenitors leading to inhibition of erythropoiesis resulting in anemia and reticulocytopenia [42], while myeloid and megakeryocytes in the BM are usually non permissive to the virus[1]. Also, data had shown that although erythrocytes are predominately affected with presentation often of PRCA; concurrent thrombocytopenia, neutropenia or pancytopenia is found infrequently especially in patients with functioning spleen [5, 60].

However, other authors suggested that neutropenia may be either due to that circulating neutrophils could carry replicating parvovirus in acutely infected patients or may originate from an autoimmune process as shown by the decrease in cloning efficiency observed when granulocytic progenitors derived from cells harvested from BM of children with PB19-induced chronic neutropenia are cultured in vitro in autologus plasma [59].

Primary autoimmune neutropenia occurs predominantly in infancy. The BM of children with neutropenia was examined for PB19 DNA. Results indicated that PB19 infection may be a common cause of immune-mediated neutropenia in childhood (15 of 19 patients) [1].

Thus, children with hematological disorders presented with anemia, neutropenia and lymphocytosis should be evaluated for PB19 infection by the aid of laboratory methods.

Congenital Anaemia

Despite frequent reports of intrauterine PB19 infection and hydrops, the risk of associated congenital anemia is apparently low. This may be explained by PB19 causing severe disease in only the first two trimesters [61]. However, intrauterine infection by PB19 in immunocompromised fetus can lead to severe congenital aplastic anaemia [62].

Data had reported a case of intrauterine infection by human PB19 manifested as ascitis during pregnancy, at delivery, a pale female infant with bradycardia was delivered. PCR lab tests revealed that the mother and the fetus were infected by PB19. The neonate was born with low haemoglobin (Hb = 10gm/dl) and with ascitis [64].

Recently, PB19-induced congenital anemia is corrected by multiple intrauterine transfusions and postnatal immunoglobulin [64].

Haemophagocytic Syndrome

Haemophagocytic syndrome or haemophagocytic lymphohistiocytosis is a rare disease that is often fatal despite treatment. Haemophagocytic syndrome is caused by a dysregulation in natural killer T-cell function, resulting in activation and proliferation of lymphocytes or histiocytes with uncontrolled haemophagocytosis and cytokine overproduction. The syndrome is characterised by fever, hepatosplenomegaly, cytopenias, liver dysfunction, and hyperferritinaemia. Haemophagocytic syndrome can be either primary, with a genetic aetiology, or secondary, associated with malignancies, autoimmune diseases, or infections. Infections associated with haemophagocytic syndrome are most frequently caused by viruses, particularly Epstein-Barr virus (EBV) [65].

PB19 is one of several viruses that have been implicated as causing viral-associated hemophagocytic syndrome (VAHS). Thus VAHS is generally considered a non-specific response to a variety of viral insults rather than a specific manifestation of a single pathogen. It is not uncommon syndrome and occurs in a wide rang of infections, not only viral, but also in the context of bacterial, rickettsial, fungal, and parasitic infections [1].

Haemophagocytic is characterized by histiocytic hyperplasia, marked hemophagocytosis, and cytopenia, in association with a systemic viral illness. The term hemophagocytosis describes the pathologic finding of activated macrophages, engulfing erythrocytes, leukocytes, platelets, and their precursor cells in the BM and other tissues. In contrast to malignant histiocytosis, VAHS is usually a benign self-limiting illness, in which histiocytic proliferation is reversible [66].

The clinical course of haemophagocytic can be dramatic and the prognosis is often poor. The pathogenesis of VAHS is not well understood. Many believe that viral infection provokes an abnormal immune response in predisposed individuals leading to hyperactivation of Th1 helper cells, macrophage proliferation and secretion of large amounts of cytokines. The resultant hypercytokinaemia may be responsible for the clinical and biochemical manifestations of haemophagocytic [67].

Many cases of PB19– associated hemophagocytic syndrome have been reported. The first case was reported in1990; an acute febrile illness with pancytopenia developed in a previously healthy child. His bone marrow showed hemophagocytosis, and the diagnosis of PB19 infection was documented serologically by the presence of Bl9-specific IgM and IgG antibodies. Splenectomy was performed for diagnostic purposes, pancytopenia resolved immediately following splenectomy and the patient recovered completely [68].

The syndrome can be associated with other hematological disorders. Four cases occurred in patients with underlying hematologic problems. One with acute lymphoblastic leukemia (ALL) presented with unexplained neutropenia, his BM examination revealed histiocytic infiltration and hemophagocytosis. Diagnosis was confirmed by demonstrating PB19 DNA in the serum. Recovery followed a temporary cessation of chemotherapy [69].

In another patient with Evan's syndrome, aplastic crisis and profound thrombocytopenia developed. His BM examination revealed hemophagocytosis without any disturbance in megakaryopoiesis. The diagnosis was confirmed by demonstrating PB19 DNA and specific IgM and IgG antibodies in the serum and the patient recovered spontaneously within 4 weeks [70].

The other two cases occurred in patients with hereditary spherocytosis; both of them had a transient acute febrile illness with pancytopenia, reticulocytopenia, lymphadenopathy, and splenomegaly. PB19 DNA was documented in their sera [71]. The investigator suggested that hemophagocytosis associated with PB19 infection is underestimated and that hemophagocytosis could explain the commonly found neutropenia and thrombocytopenia associated with PB19 infection.

In Japan, a research was between August 1992 and August 1994 found seven adult patients with hemophagocytic syndrome associated with PB19 infection. Five of the patients were previously healthy but the other two had underlying immunosuppressive disease; the former five patients recovered spontaneously without any treatment. Cases of PB19-associated hemophagocytic syndrome have been reported; usually children and the with immunosuppressive disease or hematological disorder. These data suggests that PB19 is not a rare cause of virus-associated hemophagocytic syndrome (VAHS). Moreover, PB19-associated hemophagocytic syndrome could occur more frequently than previously thought, not only in children and adults with underlying disease but also in adults, even in good health [72].

Chronic Bone Marrow Failure due to Persistent PB19 Infection

A lack of protective antibodies allows PB19 to persist. In the absence of antiviral immunity, fifth disease does not develop (because antigen–antibody complexes are not formed), but pure red-cell aplasia (PRCA) can be a manifestation of persistent PB19 infection As a result of that, chronic BM failure develops due to failure of the immune system to mount an effective response to the virus.

The anemia is severe and requires transfusion; reticulocytes are absent from the blood, as are erythroid precursors from the marrow; giant pronormoblasts in a congruous clinical setting may lead to the diagnosis. Antibodies to parvovirus are usually absent; however, the virus can be readily detected in the circulation, often at extremely high levels (>10^{12} genome copies per milliliter) [37].

Bone marrow transplantation (BMT) is recommended for severe aplastic anemia in a patient with persistent human PB19 infection. This is a case report of 9-year-old girl with no previous history of immunodeficiency that developed severe aplastic anemia concurrent with PB19 persistent infection. Both IgM antibody to PB19 and PB19 DNA identified by real-time PCR were found in the patient's serum at time of diagnosis of aplastic anemia. No giant proerythroblasts were found in her bone marrow at diagnosis [73].

Although intravenous immunoglobulin (Ig) reduced serum PB19 DNA, the aplastic status of her bone marrow did not improve. Both aplastic anemia and persistent PB19 viremia were successfully treated by BMT from an HLA-identical sibling donor. Serum PB19 DNA increased temporarily after BMT; however, neither organ nor marrow failure was observed. PB19 DNA disappeared from the serum 2 months after BMT, suggesting that a normal immune response was restored by BMT and terminated the PB19 viremia. During BMT, use of high-titer intravenous Ig for PB19 might prevent PB19-associated organ failure [73].

Role of PB19 in Acute Leukemia (AL)

Trials have been made to explore the etiology for development of acute leukemia. Infections are believed to play a major role in development of acute leukemia especially in children. There is debate from studies on early childhood infectious exposures, vaccination and social mixing. Some supportive evidence for an infectious etiology supported by the findings of space-time clustering and seasonal variation of acute leukemia new cases presentations.

Generally speaking, there are three current hypotheses concerning infectious mechanisms in the etiology of childhood leukemia: either in utero exposure or around the time of birth, delayed exposure beyond the first year of life to common infections and unusual population mixing. Till now, no specific virus has been definitively linked with childhood leukemia and there is no evidence to date of viral genomic inclusions within leukemia cells. Nevertheless, case–control and cohort studies have revealed equivocal results. Maternal infection during pregnancy has been linked with increased risk whilst protective roles were determined to be breast feeding and day care attendance in the first year of life with immune state development to various infectious agents [74].

Spatial clustering suggests that higher incidence is associated with specific areas with increased levels of population mixing, particularly in previously isolated populations which carry risk of acquiring infections. Ecological studies have also shown excess incidence with higher population mixing.

The marked childhood peak in resource-rich countries and an increased incidence of the childhood peak in acute lymphoblastic leukemia (ALL) (occurring at ages 2–6 years predominantly with precursor B-cell ALL) is supportive of the concept that reduced exposure to early infection may play a role.

In addition, genetically determined individual response to infectious agents may be critical in the proliferation of preleukemic clones as evidenced by the human leucocyte antigen class II polymorphic variant association with precursor B-cell and T-cell ALL [74].

Previous data led by Greaves (1999) has demonstrated that for certain types of acute leukemia as infant ALL with MLL rearrangements, mostly in the precursor B-cell ALL (certainlyTEL-AML1 and hyperdiploid ALL) and childhood AML with t(8;21) translocations, the first genetic event in childhood leukemia involves production of a preleukaemic clone in utero. It is the postnatal events including infection are almost certainly required to promote the development of clinical leukemia with the gene rearrangements in utero.

These data has clearly defined the need for at least two and possibly multiple postnatal events. Therefore, when considering etiology the initiating events occurring most frequently in utero and subsequent events occurring postnatally are important. Whether infection plays a part at either or both time-points is unclear.

It is claimed that the first 'genetic' event might well be initiated by infection. However, all the studies to date have shown no evidence of viral genomic inclusion in leukemia cells. Thus, the suggestion that maternal infection during pregnancy may be associated with an increased risk of childhood leukemia would be supportive of such a relationship [35].

The hypothesis includes the possibility that infection might occur in utero or near to birth. The space time clustering data based on time and place of birth would clearly be consistent with events in utero [76, 77].

The findings from space-time clustering studies from the UK and Sweden [74, 78] support the in utero exposure, clearly in the most recent time period.

Obviously there is quite a lot of supportive data to suggest that the later postnatal events are indeed related to infections and/or the body's response to them. Evidence from the geographical incidence and temporal increases in the incidence of childhood leukemia is consistent with lack of immune stimulation during the first few years of life, particularly in more affluent communities and societies.

The limited and protective effect of breast feeding and more significantly, the apparent benefit of increased social contact because of nursery attendance in the first year of life would again be supportive of the concept that leukemia is more likely if there is an absence of infectious exposure early in life and that leukemia results from some form of delayed and possibly abnormal response to infection at a later age.

The host factor represented by HLA class II allele linkage to precursor B-cell ALL is further corroborative evidence [79]. Space-time clustering studies centered on time and place of diagnosis are also consistent with the delayed infection hypotheses, as are those based on time of diagnosis and place of birth and also Kinlen's studies on unusual population mixing.

It is very important to realize that the population mixing hypotheses are not mutually exclusive. Elements of both may be involved in individual cases. Infection may initiate a preleukemic clone but that may not lead to overt leukemia in the absence of delayed later infection and/or abnormal response to such infection. Controversially to benefits of population mixing this may increase the chance of infectious exposure in susceptible people at any stage.

At the present time, in the absence of definitive evidence regarding either specific single or multiple infectious culpable agents at either time-point, it is beneficial to continue to investigate events and exposures around the two time-points and the molecular events that result from such exposures.

Greater clarification as to likely etiological agents may emerge. For example, in infant leukemia, where very characteristic rearrangement of the MNLL gene occurs, there has been both epidemiological and molecular evidence to suggest that exposure to naturally occurring topoisomerase II inhibitors and the inability by the mother and fetus to metabolize them rapidly significantly increases the risk of developing leukemia.

Topoisomerase II reduces DNA damage that is inappropriately repaired. A combination of exposure, DNA damage and inappropriate repair plus failure to metabolize the inhibitors all contributes to the initiation of a rare leukemia.

So we can say that there may be a multiple pathways involved in the development of any childhood leukemia that will always involve initial breakage and inaccurate repair of DNA in response to infection, chemicals, low level irradiation or other, as yet unknown, environmental exposures. Then one or more subsequent events convert a preleukemic clone into an overt malignant population. Two genetic events and possibly also proliferative stimuli and/or suppression of bone marrow are all required to produce an overt leukemia.

Another important issue is the genetic susceptibility. It looks likely in terms of not only the response to infection, but also the recognition and repair of DNA.

The study of the molecular events associated with leukemia transformation may help us to better understand the changes.

To date, evidence cannot confirm or refute whether at least one infection induces or is a major co-factor for developing ALL or cALL, or perhaps actually protects against disease. Differences in methodology and populations studied may explain some inconsistencies. Other challenges to proof include the likely time lag between infection and diagnosis, the ubiquity of many infections, the influence of age at infection, and the limitations in laboratory assays; small numbers of cases, inaccurate background leukemia rates, and difficulty tracking mobile populations further affect cluster investigations. Nevertheless, existing evidence partially supports plausibility and warrants further investigation into potential infectious determinants of ALL and cALL, particularly in the context of multifactorial or complex systems [80].

Therefore, at present we are still some time away from measures that would enable us to subtly target therapy or indeed apply preventative measures based on a clear understanding of causation, including immune modulation of response to infection.

One of the implicated viruses in development of acute leukemia is PB19. It is suggested that the interaction between PB19 and host immune response has a role. Acute parvovirus infection is associated with a significant cytokine cascade, which is associated with a degree of disturbed haematopoiesis and/or suppression of normal marrow function, which may allow 'release' of low level malignant clones or indeed, induce acute leukemia. Serum concentrations of TNFα and IFNγ are typical of symptomatic acute PB19 infection [81]. Greatly raised serum concentrations of interleukin 6 (IL-6) and granulocyte monocytes colony stimulating factor (GM-CSF) have previously been documented at the onset of acute leukemia [82, 83].The chemokine MCP-1 is released by leukemia cells; it is chemotactic for monocytes and induces the tumoricidal activity of monocytes, and is thus a possible host defense against neoplasia [84]. PB19 virus infection is associated with the suppression of erythroid elements in the bone marrow, along with immune cell proliferation and up regulation of key mediators, such as GM-CSF. Such mechanisms may play an important role in the conversion of preleukaemic clones to an overt leukemia.

Another important factor in this event is association of PB19 infection with certain HLA types. Data has shown that 94% of patients with acute symptomatic parvovirus B19 carry HLA-DRB1 alleles with a neutrally charged glutamine (Q) at position 10 and a positively charged lysine (K) at position 12 within the first hypervariable region [85]. This includes HLA-DRB1 alleles *01, *04, *07, *09, *15 and *16. Polymorphism in the first hypervariable region of the DRB1 molecule has been associated with other clinical conditions, [86,87] and we have speculated that it may be a crucial region in the immune response to parvovirus B19.

It is interesting that each of the four patients in one study carried at least one allele with the same polymorphism in this region. Although only a small number, it provides further support to the hypothesis that the immune response is an important factor in the pathogenesis of PB19 associated acute leukemia [88].

Greaves has long postulated that infection plays a part in childhood ALL [89]. Epidemiological (for example, population mixing, space time clustering, seasonality, etc) and biological (for example, association with particular HLA alleles) data suggest a role for infectious agents at some stage in the etiology of childhood ALL. Other studies have suggested in utero initiation of preleukaemic clones,[90] which require one or more subsequent genetic and/or proliferative event(s) to produce the leukemia—certainly in ALL, and maybe also in AML, during early childhood. Although it is possible that infection may play a role in initiation of the initial genetic rearrangements, there is increasing evidence that its major role lies at the later proliferative stage. The host response to infection may play an additional role in allowing proliferation of the premalignant clone.

The host response during PB19 associated acute leukemia may provide insights into possible mechanisms by which the virus may at least precipitate the overt leukemia.

The authors described many roles of PB19 in AL, some hypothesized that it is a risk factor in developing acute leukemia (AL) & others hypothesized that it has potential effects to precipitate different forms of cytopenia in patients prior to or at the diagnosis of AL [91]. There may be an association between AL and PB19 infection which results from immunosupression as reported by more than 20 reports [92]. It seems also that repeated blood transfusions to those patients could play a role in such high prevalence of PB19.

Nevertheless, several reports describe PB19 infection preceding and mimicking ALL [93]. Studies have suggested in utero initiation of preleukaemic clones which require one or more subsequent genetic and/or proliferative event(s) to produce leukemia, certainly in ALL, and may be also in AML, during early childhood. Although it is possible that infection may play a role in initiation of the initial genetic rearrangements, there is increasing evidence that its major role lies at the later proliferative stage. The host response to infection may play an additional role in allowing proliferation of the premalignant clone [88].

The host response during PB19 associated acute leukemia may provide insights into possible mechanisms by which the virus may at least precipitate the overt leukemia. PB19 virus infection is associated with the suppression of erythroid elements in the BM, along with immune cell proliferation and upregulation of key mediators, such as granulocyte–macrophage colony stimulating factor (GM-CSF). Also serum concentrations of tumor necrosis factor-α (TNFα) and γ-inteferon (IFNγ) which are typical of symptomatic acute PB19 infection, greatly raised serum concentrations of interleukin-6 (IL-6) and GM-CSF have previously been documented at the onset of acute leukaemia. Such mechanisms may play an important role in the conversion of preleukaemic clones to an overt leukaemia [88].

According to these authors who hypothesised that PB19 virus does play a role in the pathogenesis of acute leukemia, the virus may no longer be present in the serum at the time of diagnosis, but may be present at other more cryptic sites such as cerebrospinal fluid (CSF) because clearance from these areas may be delayed [88].

The authors present two cases of PB19 infection and BM infiltration with pre-B-cell lymphoblasts; one patient was diagnosed as having ALL, and the other patient, with neurologic findings, showed total resolution of the blastic morphology and phenotype [91].

On the other hand, authors found that PB19 infection may infrequently serve as a prodrome to ALL. A preleukaemic phase, typified by pancytopenia and BM hypoplasia, is an uncommon but well-documented prelude to ALL (pre-ALL) in children. PB19 exhibits a marked tropism to human BM and replicates only in erythroid progenitor cells acting as a confounding, but treatable agent in immunocompromised patients.

Heegaard & his colleagues (2001).presented the first case of PB19-associated pre-ALL characterized by severe and recurring transient pancytopenia in a child who developed ALL 5 months later. The advent of PB19-specific IgG at the time of infection and the subsequent disappearance 1.5 years later has not previously been described. In this patient the observed cytopenias were probably the result of PB19 acting in concert with the failing BM and PB19 is possibly one of several factors capable of triggering the onset of pre-ALL. Recent data gives a clue that there is possible correlation between P15 (INK4B) promoter methylation and parvovirus B19 infection is observed in adult acute leukemia patients [95].

There has been much speculation about the role of PB19 in childhood AL; some hypothesized that it is a risk factor in developing AL which has been suggested, on the basis of findings in epidemiological studies, that AL may be initiated by an in utero infection of the fetus [96]. Others hypothesized that it has potential effects to precipitate different forms of cytopenia in patient's proir to or at the diagnosis of AL [91].

In contrary; other authors found that PB19 DNA was not determined in any of the umbilical cord blood from children who later developed ALL or in the healthy controls. These findings suggest that it is less likely that childhood ALL is associated with an in utero infection with PB19 [97].

A recent report pointed out the expression of PB19 receptor on leukemic cells with weak association [98]. Alternatively, PB19 infection in a patient with subclinical immunodeficiency might have led to the development of leukemia or PB19 infection may contribute to clonal proliferation [91].

The presenting symptoms of PB19 infection associated with leukemia differ in patients. Fattet et al., (2004) reported PB19 infection in five patients treated for ALL. The presenting symptoms suggesting PB19 infection were pallor, fatigue, petechiae and pancytopenia in four patients and generalized rash in two patients. There are no study to confirm the association between PB19 infection and HSM and lymphadenopathy in leukemic patients, so the presence of these two signs with high rate in patients with acute leukemia are often due to extramedullry hemopoiesis and leukemic infiltration present with AL regardless the presence or absence of PB19 infection [100].

El-Mahallawy et al., (2004) studied ALL patients who were receiving maintenance chemotherapy for the presence of PB19 virological markers. It was found that PB19 infection was accompanied by low Hb (mean= 7.8 ± 1.5 g/dl), normal white blood cells count (WBCs count) (mean= $3.5 \pm 2.3 \times 10^3$ / mm^3) except four patients were neutropenic with WBCs count ($< 1 \times 10^3$ / mm^3) and normal platelet counts (mean= 213.5×10^3 / mm^3).

Also, Fattet et al., 2004 reviewed that the infection occurred generally after several months of immunosuppression with non-specific clinical signs, but hematologic manifestations (anemia, thrombocytopenia) were often described, pancytopenia was not a predominant hematologic manifestation.

Heegaard and Schmiegelow (2002) reviewed that PB19 infection was able to mimic a leukemic relapse or therapy induced cytopenia in ALL patients. It was reported that PB19 viremia may persists up to 10 months in patient's sera [103] and BM [104] of immunocompetent individuals, if such individual develops a medication-induced immunodeficiency; suppressed erythropoiesis was observed as evidenced by reduced BM normoblasts [101].

Yetgin et al., (2004) speculated that the role of PB19 in recently diagnosed AL still unclear, it may induce leucocytosis and lymphocytosis with normal Hb and platelets in some patients, while leucopenia, neutropenia and lymphocytosis in the other patients.

So, although not clearly defined, an etiologic role of viruses and PB19 infection in the development of leukemia and/or ALL is increasingly being suggested [93]

In contrary, Heegaard et al., (1999) concluded that pediatric ALL patients prior to or at the time of diagnosis are not particularly susceptible to PB19 infection and the associated clinical significant pancytopenia is rare. However, PB19 infection may infrequently serve as a prodrome to ALL. Furthermore studies are needed to clarify the role of PB19 in the occurrence of AL.

Laboratory diagnosis of PB19 in those patients is a deilemma. El-Mahallawy et al. (2004) who concluded that dependence on serology testing only would have missed the diagnosis of PB19 infection in some patients. On the other hand; some patients showed demonstrable IgM without the presence of DNA, so depending on DNA testing alone would have missed the diagnosis in these cases.

Similarly, comparison of PCR and serologic results indicated that serologic testing alone is insufficient for the diagnosis of PB19 infections in immunocompromized patients [105].

Fattet et al. (2004) concluded that the absence of PB19 IgM during acute infection is not uncommon in patients with ALL. The host's immune state is an important factor determining the immune response to viral infection. In immunocompetent individuals; laboratory diagnosis of PB19 infection can be built on the basis of serologic testing (IgM & IgG) antibodies, while DNA assays may be required in immunocompromised patients who are unable to amount an adequate antibody response [1].

Other data indicate that the number of children with immune response to PB19 proteins is distinctly higher than the number of children with PB19 DNA. However, there was establishment and maintenance of a persistent

PB19 infection with PB19 DNA in bone marrow. This demonstrates that there may be an aberrant immune response to PB19 in patients with acute leukemia as the presence of neutralizing antibodies to PB19 did not clear the virus. Therefore, there may be defects in specific immunity and persistence of viral DNA as abnormal findings for the disease course. High levels of IgM and IgG were associated with the persistence of the virus and not the clearance of it [106].

So, it is clear that IgM for PB19 can be used as a guide for recent infection in immunocompetent patients, while the matter is different in patients with leukemia in whom molecular techniques as PCR are essential for diagnosis of PB19 infection.

Children with ALL who were infected with parvovirus B19 became cytopenic, leading to reduced treatment intensity and to complications during treatment. Screening for parvovirus B19 DNA by quantitative polymerase chain reaction in pediatric patients with ALL and unexplained cytopenia is suggested [107, 108]. Parvovirus B19 is considered a common viral infection in children with acute lymphoblastic leukemia receiving chemotherapy. We must use our full potential to exclude such infection, which can mimic the side effects of chemotherapy in these patients. In immunocompromised patients, there are immunologic discrepancies in humoral immune responses for both IgM and IgG between individuals with parvovirus B19 persistence and healthy individuals, findings that may reflect both failed immunity and antigenic exhaustion. The contemporaneous determination of parvovirus B19 DNA by PCR in both bone marrow and peripheral blood and specific serologic markers appears to be the most appropriate diagnostic protocol for the correct laboratory diagnosis of parvovirus B19 infection in these patients.

Conclusion

In conclusion, PB19 infection is detected in high rates among children with chronic hematological disorders. PB19 must be suspected and screened for when there is chronic anemia in those patients associated with neutropenia and lymphocytosis. Leukemia patients under chemotherapy presented with anemia, neutropenia and lymphocytosis need to be screened for Parvovirus before change of chemotherapy regimen.

REFERENCES

[1] Heegaard ED and Brown KE. Human Parvovirus B19. *Clin. Microb. Rev.* 2002, 15:485-505.

[2] Owren PA. Congenital hemolytic jaundice: the pathogenesis of the `hemolytic crisis'. *Blood.* 1948; 3: 231-248.

[3] Young N. Hematologic and hematopoietic consequences of B19 parvovirus infection. *Seminars in Hematology.* 1988; 25: 159±172.

[4] Chisaka H, Morita E, Yaegashi N, et al. Parvovirus B19 and the pathogenesis of anemia. *Rev. Med. Virol.* 2003, 13:347-59.

[5] Young NS, Brown KE. Parvovirus B19. *NEJM.* 2004, 350, 586-597

[6] Brown KE. Hematological consequences of Parvovirus B19 infection. *Baillieres Best. Pract. Res. Clin. Hematol.* 2000, 13:245-59.

[7] Brown KE, Anderson SM, Young NS. Erythrocyte P antigen: cellular receptor for B19 parvovirus. *Science.* 1993;262: 114-117.

[8] Wagner AD, Goronzy JJ, Matteson EL, Weyand CM. Systemic monocyte and T cell activation in a patient with human parvovirus B19 infection. *Mayo Clin. Proc.* 1995; 70:261-265.

[9] Murai C, Munakata Y, Takahashi Y, et al. Rheumatoid arthritis after human parvovirus B19 infection. *Ann. Rheum. Dis.* 1999; 58:130-132.

[10] Anderson MJ, Higgins PG, Davis LR, et al. Experimental parvovirus infection in humans. *J. Infect. Dis.* 1985; 152:257-265.

[11] Barlow GD, Mckendrick MW. Parvovirus B19 causing leucopenia and neutropenia in a healthy adult. *J. Infect.* 2000; 40:192-195.

[12] Nesher G, Osborn TG, Moore TL. Parvovirus infection mimicking systemic lupus erythematosus. *Semin. Arthritis. Rheum.* 1995;24: 297-303.

[13] Kerr JR, Barah F, Mattey DL, et al. Circulating tumor necrosis factor-α and interferon–γ are detectable during acute and convalent parvovirus B19 infection and are associated with prolonged and chronic fatigue. *J. Gen. Virol.* 2001;82: 3011-3019.

[14] Weigel-Kelly KA, Yoder MC, Srivastava A. Recombinant human parvovirus B19 vectors: erythrocyte P antigen is necessary but not sufficient for successful transduction of human hematopoietic cells. *J. Virol.* 2001;75 :4110-4116.

[15] Kaufmann B, Baxa U, Chipman PR, Rossmann MG, Modrow S, Seckler R. Parvovirus B19 does not bind to membrane-associated globoside in vitro. *Virology.* 2005; 332:189-198.

[16] Weigel-Kelley KA, Yoder MC, Srivastava A. α5β1 integrin as a cellular co-receptor for human parvovirus B19: requirement of functional activation of {beta} 1 integrin for viral entry. *Blood.* 2003; 102:3927-3933.

[17] Mimori T, Ohosone Y, Hama N, et al. Isolation and characterization of cDNA encoding the 80-kDa subunit protein of the human autoantigen Ku(p70/p80) recognized by autoantibodies from patients with scleroderma-polymyositis overlap syndrome. *Proc. Natl. Acad. Sci. U.S.A.* 1990;87: 1777-1781.

[18] Adam L, Bandyopadhyay D, Kumar R. Interferon-α signaling promotes nucleus-to-cytoplasmic redistribution of p95Vav, and formation of a multisubunit complex involving Vav, Ku80, and Trk-2. *Biochem. Biophys. Res. Commun.* 2000; 267: 692-696.

[19] Hector L, Darren G, Guillermo ET. Novel localization of the DNA-PK complex in lipid rafts. *J. Biol. Chem.* 2003; 278:22136-22143.

[20] Prabhakar BS, Allaway GP, Srinivasappa J, Notkins AL. Cell surface expression of the 70-kD component of Ku, a DNA-binding nuclear autoantigen. *J. Clin. Invest.* 1990;86:1301-1305.

[21] Morio T, Hanissian SH, Bacharier LB, et al. Ku in the cytoplasm associates with CD40 in human B cells and translocates into the nucleus following incubation with IL-4 and anti-CD40 mAb. *Immunity.* 1999;11:339-348.

[22] Monferran S, Paupert J, Dauvillier S, Salles B, Muller C. The membrane form of the DNA repair protein Ku interacts at the cell surface with metalloproteinase 9. *EMBO J.* 2004; 23: 3758-3768.

[23] Sylvie M, Catherine M, Lionel M, Philippe F, Bernard S. The membrane-associated form of the DNA repair protein Ku is involved in cell adhesion to fibronectin. *J. Mol. Biol.* 2004; 26:503-511.

[24] Fu Y, Ishii KK, Munakata Y, Saito T, Kaku M, Sasaki T. Regulation of tumor necrosis factor α promoter by human parvoviris B19 NS1 through activation of AP1 and AP2. *J. Virol.* 2002; 76:5359-5365.

[25] Lynch EM, Moreland RB, Ginis I, Perrine SP, Faller DV. Hypoxia-activated ligand HAL1/13 is lupus autoantigen Ku80 and mediates lymphoid cell adhesion in vitro. *Am. J. Physiol. Cell Physiol.* 2001; 280:897-911.

[26] Morey AL, Ferguson DJ, Fleming KA. Ultrastructural features of fetal erythroid precursors infected with parvovirus B19 in vitro: evidence of cell death by apoptosis. *J. Pathol.* 1993; 169:213–20.

[27] Sol N, Le Junter J, Vassias I, Freyssinier JM, Thomas A, Prigent AF, *et al*. Possible interactions between the NS-1 protein and tumour necrosis factor alpha pathways in erythroid cell apoptosis induced by human parvovirus B19. *J. Virol.* 1999;73: 8762–70.

[28] Kellermayer R, Faden H and Grossi M. Clinical presentation of Parvovirus B19 infection in children with aplastic crisis. *The Pediatric Infectious Disease J.* 2003, 22:1100-1.

[29] Pallier C, Greco A, Le Junter J, Saib A, Vassias I, Morinet F. The 3¢ untranslated region of the B19 parvovirus capsid protein mRNAs inhibits its own mRNA translation in nonpermissive cells. *J. Virol.* 1997; 71: 9482–9.

[30] Qian XH, Zhang GC, Jiao XY, et al. Aplastic anemia associated with Parvovirus B19 infection. *Arch. of Dise. in Child.* 2002. 87:436-37.

[31] Smith-Whitley K, Zhao H, Hodinka RL, et al. Epidemiology of human parvovirus B19 in children with sickle cell disease. *Blood.* 2004:, 103: 422-27.

[32] Rao SP, Miller ST and Cohen BJ. Transient aplastic crisis in patients with sickle cell disease. B19 parvovirus studies during a 7-year period. *Am. J. Dis. Child.* 1992, 146:1328-30.

[33] Brown KE and Young NS. Parvoviruses & Bone Marrow Failure. *Stem cells.* 1996,14:151-63.

[34] Barlow GD, McKendrick MW. Parvovirus B19 causing leucopenia and neutropenia in a healthy adult. *J. Infect.* 2000; 40:192-195.

[35] Smith JC, Megason GC, Iyer RV, et al. Clinical characteristics of children with hereditary hemolytic anemias and aplastic crisis: A 7- year review. *South. Med. J.* 1994, 87:702-7.

[36] Zimmerman SA, Davis JS, Schultz WH, et al. Subclinical parvovirus B19 infection in children with sickle cell anemia. *J. Pediatr. Hematol. Oncol.* 2003, 25:387-9.

[37] Brown KE, Anderson SM, Young NS. Erythrocyte P antigen: cellular receptor for B19 parvovirus. *Nature.*1993, 262, 114-117.

[38] Hasle H, Kerndrup G, Jacobsen BB, et al. Chronic parvovirus infection mimicking myelodysplastic syndrome in a child with subclinical immunodeficiency. *Am. J. Pediatr. Hematol. Oncol.* 1994, 16:329-33.

[39] Ozunel L, Akgun Y and Gulbas Z. The investigation of parvovirus B19 infection in patients with haematologic disorders and chronic anaemia by using PCR and ELISA. *Clin. Microb. Infect. Supplement.* 2004, 10:343-44.

[40] Nguyen QT, Sifer C, Schneider V, et al. Novel Human Erythrovirus Associated with Transient Aplastic Anemia. *J. Clin. Microb.* 1999, 37: 2483-87.

[41] Chisaka H, Morita E, Yaegashi N, et al. Parvovirus B19 and the pathogenesis of anemia. *Rev. Med. Virol.* . 2003, 13:347-59

[42] Morita E, Nakashima A, Asao H, et al. Human Parvovirus B19 Non Structural Protein (NSI) Induces Cell Cycle Arrest at G1 Phase. *J. Virology.* 2003, 77:2915-21.

[43] Wildig J; Michon P; Siba P; Mellombo M; Ura A; Mueller I; Cossart Y. Parvovirus b19 infection contributes to severe anemia in young children in papua new Guinea. *J. Infect. Dis.* 2006 15;194:2/146-53.

[44] Kellermayer R, Howard F, Mauro G. Clinical presentation of parvovirus B19 infection in children with aplastic crisis. *The Pediat. Infect. Dis. J.* 2003, 22: 12/ 1100-1101

[45] Brown KE, Young NS, Alving BM, et al. Parvovirus B19: implications for transfusion medicine. Summary of a workshop Transfusion. 2001,41:130-35.

[46] Cunningham D and Rennels MB Parvovirus B19 infection. 2002, On web site: http://www.emedicine.com/ped/topic192.htm

[47] Sabella C and Goldfarb J. Parvovirus B19 Infections. *Am. Fam. Physician.* 1999, 60:1455-60.

[48] Barash J, Dushnitzky D, Sthoeger D, et al. Human parvovirus B19 infection in children: uncommon clinical presentations. *Med. Assoc. J.* 2002, 4:763-5.

[49] Schick P. Hemolytic Anemia. 2005, On web site: www.eMedicine.com, Inc.

[50] Oliveira SA, Camacho LA, Pereira AC, et al. Clinical and epidemiological aspects of human parvovirus B19 infection in an urban area in Brazil. *Mem. Inst. Oswaldo Cruz.* 2002, 97:965-70.

[51] Heegaard ED, Mayer J, Hornsleth A, et al. Parvovirus B19 infection in patients with chronic anemia. *Haematologica.* 1997, 82: 402-5.

[52] Mishra B, Malhotra P, Ratho RK et al. Human parvovirus B19 in patients with aplastic anemia. *Am. J. Hematol.* 2005, 79:166-7.

[53] Badr MA. Human parvovirus B19 infection among patients with chronic blood disorders. *Saudi. Med. J.* 2002, 23: 295-97.

[54] Sant'Anna AL, Garcia Rde C, Marzoche M, et al. Study of chronic hemolytic anaemia patients in Rio de Janeiro: prevalence of anti-human parvovirus B19 IgG antibodies and the development aplastic crises. *Rev. Inst. Med. Trop. Sao Paulo.* 2002, 44:187-90.

[55] Parekh S, Perez A, Yang XY, Billett H. Chronic parvovirus infection and G6PD deficiency masquerading as Diamond-Blackfan anemia. *Am. J. Hematol.* 2005; 79:1/54-7.

[56] Crook TW, Rogers BB, McFarland RD, et al. Unusual BM manifestations of parvovirus B19 infection in immunocompromised patients. *Hum. Pathol.* 2000, 31: 161-8.

[57] Canavan BF, Huhn RD, Kintt C, et al. Concurrent presentation of erythrocytic and megakaryocytic aplasia. *Am. J. Haematol.* 1996, 51: 68-72.

[58] Heegaard ED, Rosthoj S, Petersen BL, et al. Role of parvovirus B19 infection in childhood idiopathic thrombocytopenic purpura. *Acta Paediatr.* 1999,88:614-17.

[59] Gautier E, Bourhis JH, Bayle C, et al. Parvovirus B19 associated neutropenia. Treatment with Rh G-CSF. *Hematol. Cell Ther.* 1997, 39:85-87.

[60] Osaki M, Matsubara K, Iwasaki T, et al. Severe aplastic anemia associated with human parvovirus B 19 infection in a patient without underlying disease. *Ann. Hematol.* 1999, 78:83–6.

[61] Yaegashi N, Niinuma T, Chisaka H, et al. Serologic study of human parvovirus B19 infection in pregnancy in Japan. *J. Infect.* 1999, 38:30-35.

[62] Veprekova L, Jelinek J, Zeman J. Congenital aplastic anemia caused by parvovirus B19 infection. *Cas. Lek. Cesk.* 2001,140:178-80.

[63] Giannakopoulou C, Hatzidaki E, Giannakopoulou K, et al. Congenital infection by human parvovirus B19 ascitis-anemia. *Clin. Exp. Obstet. Gynecol.* 1998, 25:92-93.

[64] Heegaard ED, Hasle H, Skibsted L, et al. Congenital anemia caused by parvovirus B19 infection. *Pediatr. Infect. Dis. J.* 2000, 19:1216-18.

[65] Rouphael NG, Talati NJ, Vaughan C, Cunningham K, Moreira R, Gould C. Infections associated with haemophagocytic syndrome. *Lancet Infect. Dis.* 2007 ; 7:12/814-22.

[66] Fisman DN. Haemophagocytic syndrome & infection. *Emerg. Infect. Dis.* 2000, 6:6/ 601-606

[67] Faurschou M, Nielsen OJ, Hansen PB, et al. Virus associated haemophagocytic syndrome. *Diagnosis & Treatment. Ugeskr. Laeger.* 1998, 160: 6198-200.

[68] Boruchoff SE, Woda BA and Pihan GA. Parvovirus B19- associated haemophagocytic syndrome. *Arch. Inten. Med.* 1990, 150:897.

[69] Koch WC, Massey G and Russel CE. Manifestations & treatment of human parvovirus B19 infection in immunocompromised patients. *J. Pediatr.* 1990, 116:355.

[70] Uike N, Miyamura T and Obama K. Parvovirus B19- associated haemophagocytosis in Evans syndrome: Aplastic crisis accompanied by severe thrombocytopenia. *Br. L. Hematol.* 1993,84: 530.

[71] Muir K, Todd WTA and Watson WH. Viral- associated haemophagocytosis with parvovirus B19-related pancytopenia. *Lancet.* 1992, 339: 1139-45.

[72] Shirono K and Tsuda H. Parvovirus B19- associated haemophagocytic syndrome in healthy adults. *Br. J. Haematol.* 1995, 89:923-26.

[73] Goto H, Ishida A, Fujii H, et al. Successful bone marrow transplantation for severe aplastic anemia in a patient with persistent human parvovirus B19 infection. *Int. J. Hematol.* 2004, 79:384-86.

[74] McNally RJ, Eden TO. An infectious aetiology for childhood acute leukaemia: a review of the evidence. *Br. J. Haematol.* 2004, 127:3, 243-63.

[75] Greaves M. Molecular genetics, natural history and the demise of childhood leukaemia. *Europ. J. Can.* 1999, 35: 1941–1953.

[76] Kinlen, L. Evidence for an infective cause of childhood leukaemia: comparison of a Scottish New Town with nuclear reprocessing sites in Britain. *Lancet.* 1988, ii, 1323–1327.

[77] Kinlen, L. Epidemiological evidence for an infective basis in childhood leukaemia. *Br. J. Can.* 1995, 71, 1–5.

[78] Gustafsson B. & Carstensen J. Space-time clustering of childhood lymphatic leukaemias and non-Hodgkin's lymphomas in Sweden. *Europ. J. Epidemiol.* 2000, 16: 1111–1116.

[79] Kerr JR, Mattey DL, Thomson W, *et al.* Association of symptomatic acute parvovirus B19 infection with HLA class I and class II alleles. *J. Infect. Dis.* 2002; 186:447–52.

[80] O'Connor SM, Boneva RS. Infectious etiologies of childhood leukemia: plausibility and challenges to proof. *Environ. Health Perspect.* 2007, 115:146-50

[81] McNally, R.J.Q., Alexander, F.E. & Birch, J.M. Space-time clustering analyses of childhood acute lymphoblastic leukemia by immunophenotype. *Br. J. Can.* 2002, 87: 513–515.

[82] Kerr JR, Barah F, Mattey DL, *et al.* Serum tumour necrosis factor-α (TNF-α) and interferon-γ (IFN-γ) are detectable during acute and

convalescent parvovirus B19 infection and are associated with prolonged and chronic fatigue. *J. Gen. Virol.* 2001; 82:3011–19.

[83] Sugiyama H, Inoue K, Ogawa H, *et al.* The expression of IL-6 and its related genes in acute leukaemia. *Leuk. Lymphoma.* 1996; 21:49–52.

[84] Elbaz O, Shaltout A. Implication of granulocyte–macrophage colony stimulating factor (GM-CSF) and interleukin-3 (IL-3) in children with acute myeloid leukaemia (AML); malignancy. *Hematology.* 2001; 5:383–8.

[85] Gerard C, Rollins BJ. Chemokines and disease. *Nat. Immunol.* 2001; 2:108–15.

[86] Kerr JR, Mattey DL, Thomson W, et al. Association of symptomatic acute parvovirus B19 infection with HLA class I and class II alleles. *J. Infect. Dis.* 2002; 186:447–52.

[87] Mierau R, Dick T, Bartz-Bazzanella P, et al. Strong association of dermatomyositis-specific Mi-2 autoantibodies with a tryptophan at position 9 of the HLA-DR beta chain. *Arthritis Rheum.* 1996; 39:868–76.

[88] Vrethem M, Ernerudh J, Cruz M, et al. Susceptibility to demyelinating polyneuropathy in plasma cell dyscrasia may be influenced by amino acid position 9 of the HLA-DR beta chain. *J. Neuroimmunol.* 1993; 43:139–44.

[89] Kerr JR, Barah F, Smith J, et al. Association of acute parvovirus B19 infection with new onset of acute lymphoblastic & myeloblastic leukaemia. *J. Clin. Pathol.* 2003, 56:873-75

[90] Greaves MF. Aetiology of acute leukaemia. *Lancet.* 1997;349:344–9.

[91] Wiemels JL, Cazzaniga G, Daniotti M, *et al.* Prenatal origin of acute lymphoblastic leukaemia in children. *Lancet.* 1999; 354:1499–503.

[92] Yetgin S, Cetin M, Aslan D, et al. Parvovirus B19 infection presenting as pre B-cell acute lymphoblastic leukaemia: a transient & progressive course in two children. *J. Pediatr. Hematol. Oncol.* 2004, 26:689-92.

[93] Mihneva Z. Clinical manifestations, risks and trends regarding human Parvovirus B19 infection. *Med. Rev.* 2005, 41:5-14.

[94] Savasan S and Ozdeir O. Parvovirus B19 infection & acute lymphoblastic leukaemia. *Br. J. Hemato.* 2003, 120:166-71.

[95] Heegaard ED, Hans O, Schmiegelow K, et al. Transient pancytopenia preceding acute lymphoblastic leukaemia (pre-ALL) precipitated by parvovirus B19. *Br. J. Haematol.* 2001,114:810-13.

[96] Yalcin A, Serin MS, Emekdas G, Tiftik N, Aslan G, Eskandari G, Tezcan S. Promoter methylation of P15 (INK4B) gene is possibly

associated with parvovirus B19 infection in adult acute leukemias. *Int. J. Lab. Hematol.* 2009 ; 31:4, 407-19..

[97] Wiemels JL, Cazzaniga G, Daniotti M, *et al.* Prenatal origin of acute lymphoblastic leukaemia in children. *Lancet.* 1999; 354:1499–503.

[98] Isa A, Priftakis P, Broliden K, et al. Human parvovirus B19 DNA is not detected in Guthrie cards from children who have developed acute lymphoblastic leukemia. *Pediatr. Blood Cancer.* 2004, 42:357-60.

[99] Cooling LLW, Zhang DS, Naides SJ, et al. Glycosphingolipid expression in acute nonlymphocytic leukemia: common expression of Shiga toxin and parvovirus B19 receptors on early myeloblasts. *Blood.* 2003, 101:711–721.

[100] Fattet S, Cassinotti P, Popovic MB. Persistent Human Parvovirus B19 Infection in Children Under Maintenance Chemotherapy for Acute Lymphocytic Leukemia. *J. Ped. Hemat. Oncol.* 2004, 26:497-503.

[101] Zaki ME and El ashry R. Clinical and hematological study for Parvovirus b19 infection in children with acute leukemia. *Int. J. Lab. Hematol.* 2009 Feb 26.

[102] El-Mahallawy HA, Mansour T, El-Din SE, et al. Parvovirus B19 infection as a cause of anaemia in pediatric acute lymphoblastic leukaemia patients during maintenance chemotherapy. *J. Pediatr. Hematol. Oncol.* 2004, 26:403-6.

[103] Heegaard ED and Schmiegelow K. Serologic study on parvovirus B19 infection in childhood acute lymphoblastic leukaemia patients during chemotherapy: clinical & hematologic implications. *J. Pediatr. Hematol. Oncol.* 2002, 24:368-73.

[104] Flunker G, Peters A, Wiersbitzky S, et al. Persistant parvovirus B19 infection in immunocompromized children. *Med. Microbiol. Immunol.* 1998, 186:189-94.

[105] Cassinotti P and Siegl G. Quantitative evidence for persistence of human parvovirus B19 DNA in an immunocompetent individual. *Eur. J. Clin. Microbiol. Infect. Dis.* 2000, 19:886-87.

[106] Schleuning M, Jager G, Holler E, et al. Human parvovirus B19-associated disease in bone marrow transplantation. *Infection.* 1999, 27:114-7.

[107] Zaki, M E S. Relevance of Specific Immunoglobulin M and Immunoglobulin G for Parvovirus B19 Diagnosis in Patients With Acute Lymphoblastic Leukemia Receiving Chemotherapy Prospective Study. *Arch. Pathol. Lab. Med.* 2007, 131, 1697-1699.

[108] Lindblom A, Heyman M, Gustafsson I, Norbeck O, Kaldensjö T, Vernby A, Henter JI, Tolfvenstam T, Broliden K. Parvovirus B19 infection in children with acute lymphoblastic leukemia is associated with cytopenia resulting in prolonged interruptions of chemotherapy. *Clin. Infect. Dis.* 2008, 15; 46:4,528-36.

Chapter 5

PB19 AND BLOOD TRANSFUSION

Concern about the safety of blood in regard to the transmission of blood-borne viruses has been increased. Safety measures have been implicated for reduction of blood transfusion associated viral infections. Combination of different strategies has been used for such purpose, such as careful selection of donors, screening for relevant virological markers and viral inactivation/removal methods. More recently, the implementation of the nucleic acid amplification technologies for the detection of HIV-1, HCV and HBV, has increased safety by reducing the "window period" of the infections. Other viruses, such as Parvovirus B19 (PB19), can cause problems for blood safety. This infection can provoke serious complications in some risk groups, such as pregnant women, patients with hematological problems, children and patients with immunodeficiency's syndromes. This chapter will discuss various aspects of this subject.

Keywords: PB19, blood transfusion, blood products, safety measures.

INTRODUCTION

A spectrum of blood-borne infectious agents is transmitted through transfusion of infected blood donated by apparently healthy and asymptomatic blood donors. The diversity of infectious agents includes hepatitis B virus (HBV), hepatitis C virus (HCV), human immunodeficiency viruses (HIV-1/2), human T-cell lymphotropic viruses (HTLV-I/II), Cytomegalovirus (CMV), Parvovirus B19, West Nile Virus (WNV), Dengue virus, trypanosomiasis, malaria, and variant CJD [1].

PB19 is normally spread via the respiratory route; parenteral transmission is possible via transfusion of blood. Blood and blood products donated during viremic state of illness carry the risk of PB19 infection. This occurs approximately 7 to 9 days after exposure and precedes the onset of symptoms by several days. Several techniques have been used to detect parvovirus B19 viremia, including an assay for circulating viral antigen [2], a hybridization assay [3-6], and more recently the more sensitive polymerase chain reaction (PCR) [6-8].

Detection of PB19 infection in blood donors had been approached via several studies. The incidence of PB19 viremia in healthy blood donors in one study was approximately 1/3,300, that is considered higher than that reported by previous surveys in which antigen detection methods were used [2] Different reports determined the prevalence of PB19 9n normal population. It was determined to be 1 in 625 blood donations (n = 16 859; range, 10^2 to $>10^7$ IU B19 DNA ml^{-1}) [9]. Previously, PB19 had been estimated to be present in 1: 16 000 transfusions, based on the mean incidence of PB19 infection in a non-epidemic period (320 cases per 100 000 population) and viraemia lasts for about 7 days [10]. During epidemics, the incidence of viraemia in donations is greatly increased, with levels as high as 1: 3790 reported in Ireland [11] and 1: 167 in Japan [11].

Other data reported incidence in blood donors approximately 1:10 000±1:25 000 units of blood during epidemic seasons will contain high titres of PB19. Incidence studies differ according to season of the study and according to the method used to detect viremia.

Screening of samples from blood donors in both USA and Germany indicate that 1:800 units contained detectable PB19 DNA by sensitive PCR (American Red Cross and Centeon, unpublished data).

The infectious level of PB19 in blood products has yet to be established with certainty and is likely to depend on the level of PB19 IgG that is co-present in the product, in addition to recipient immune status. As part of a study, a group of 100 healthy volunteers who were seronegative for PB19 were given 1 unit of plasma that had been solvent/detergent-treated [12]. Volunteers who were screened subsequently for incidences of PB19 infection, 18 % had seroconverted over the subsequent 3 months. Three of the ten batches of plasma that were used in the study were found retrospectively to contain high levels of PB19 DNA ($>10^7$ genome equivalents ml^{-1}) and these batches coincided with the plasma that was administered to the volunteers who seroconverted. Batches with low amounts of PB19 ($<10^4$ genome equivalents ml^{-1}) did not cause PB19 seroconversion. Upon discovery of these facts, the company that was involved in manufacturing this plasma (VITEX, Watertown, MA, USA) voluntarily recalled batches that were associated with B19 transmission and reviewed their screening process.

It is worth to notice that PB19 seroprevalence is higher among hemophiliacs than the general population, presumably due to the fact that products that are derived from pooled plasma are more likely to contain infectious PB19 DNA. This was shown conclusively when B19 seroprevalence in a hemophiliac population was compared with that of normal healthy individuals and it was discovered that hemophilic children who were treated with virally inactivated clotting factor had an increased level of B19 seroprevalence (92 %) [13].

Although PB19 infection may cause mild or even unrecognized illness, PB19 contamination of purified blood products can particularly be problematic as, in the absence of PB19 IgG in recipient or in immunocompromized patients; the infectious potential of B19 may be enhanced [14].

Transmission of PB19 by Blood Transfusion

It is speculated if blood transfusion can transmit PB19.It has been shown that transmission of PB19 can occur theoretically in viremic stage of infection, however the out come of infection depends upon multiple factors as viral load, the presence of neutralizing antibodies both in recipient and donor and the condition of the clinical condition of recipient .

It is noted that the possibility that PB19 virus is transmitted with single blood donations is very low, but a few cases have been documented to occur after single receive of blood. One red cell unit, contaminated by PB19 DNA, transmitted the virus to a thalassemic patient [15]. Also a case was described to develop chronic anemia due to a parvovirus B19 infection after platelet transfusion in a bone marrow transplant recipient. Another case infection in the renal transplant recipient, who developed severe anemia, was infected cadaveric renal allograft donation from a young man, in whom PBi'9 infection was diagnosed postmortem [16]. The apparent low incidence of B19 infection and disease in transfused patients might be due not only to the low coincidence of viremia in blood donors but also to coexistence of antibodies in viremic donors or in recipients. In fact, only 3 of 11 B19 DNA-positive donors had no measurable anti-B19 antibody. Susceptible recipient without protective antibodies might develop anemia consistent with PB19 infection from infected donors without antibodies [17, 18].

Transmission of PB19 by Plasma Derivatives

Blood products prepared from donated blood carry risk of PB19 transmission. Whether these blood products are infectious depends on the stability of the virus, the possible neutralization of infectivity by mixture with antibody from other units in the pool [19], and the partitioning of the virus during cryoprecipitation and cold ethanol precipitation during manufacture.

There is convincing serological evidence for transmission of B19 by non-heat-treated factor VIII and prothrombin complex concentrates [20-22].

PB19 DNA is detected in about two-thirds of the batches of non inactivated clotting-factor concentrates tested. This might be expected given the frequency of viremic donors, the size of the pools used for manufacture (3,000 to 10,000 plasma units), and the fractionation process used [23].

Though clotting factors manufacture usually sterilize factors by different methods, Parvoviruses are very heat resistant, and at high virus concentration they can withstand conventional heat treatment of blood products. Solvent detergent inactivation aimed at lipid enveloped viruses is ineffective. PB19 infection has been transmitted by steam- or dry-heated factor VIII or IX preparations and by solvent-detergent-treated factor VIII, although, in one study, hemophiliacs who received heat-treated factor VIII alone had lower prevalence of PB19 antibody and lower rates of seroconversion compared to those receiving non heat-treated factor [24,25].

Presently, plasma lots that contain high levels of PB19 are eliminated from manufacturing batches of plasma. Thus, there is a level of virus, as yet undetermined, that will not cause PB19 infection. Recent data denotes that transmission from components with $<10^{(6)}$ IU/mL viral load does not occur or, if it does, it is an uncommon event. These data do not support the need to routinely screen blood donations with a sensitive PB19V DNA nucleic acid assay.

The presence of PB19 IgG is another depending factor whether plasma pools are infectious or not. Thus, it is found that PB19 IgG levels in the range of 64.7 ± 17.5 IU ml^{-1}can be protective [12, 26,27]. Thus, it is possible that this level of PB19 IgG may be capable of preventing recipient PB19 infection when transfused with plasma that is contaminated by low levels of PB19 ($<10^4$ genome equivalents ml^{-1}).

Another blood product that carries the risk of infection is clotting factor. It has been identified in two incidences of PB19 transmission by separate lots of clotting factor concentrates and have shown that PB19 levels of 8.6×10^6

genome equivalents ml^{-1} (volume, 180 ml) and 4 x 10^3 genome equivalents ml^{-1} (volume, 966 ml) are responsible for seroconversion [14].

There have been no published reports of transmission of PB19 by immunoglobulin, albumin or plasma products although viral DNA can be detected by PCR. However, in unpublished studies, recipients of solvent-detergent plasma containing titers of PB19 DNA (107.5±108.5 genome copies/ml) did seroconvert, and in some cases this was associated with mild clinical disease. Several companies in both the USA and Europe are currently developing screening tests in an attempt to reduce the B19 viral load in blood products [28].

Clinical data on recipients concerning the infectivity of "virally inactivated" concentrates are contradictory and may reflect variation in the precise conditions used for virus removal. In one study, serological testing of hemophiliacs who had received factor VIII subjected to dry-heat treatment for 72 h at 80°C showed that they had no increased risk of infection over age-matched controls [29]. However, elevated rates of infection were found in recipients of dry heat-treated or steam-treated factor VIII [30]. Furthermore, previously non transfused hemophiliacs have been observed to become acutely infected 7 to 14 days after the first infusion of detergent-treated [31, 32].or heat-treated [33, 34] clotting factor. The persistence of

PB19 DNA in factor VIII concentrates found in this study after heat treatment for 72 h at 80°C does not necessarily indicate continued infectivity. However, these results contrast with the studies of other viruses, such as hepatitis C virus, in which the same treatment not only apparently eliminates infectivity but also completely destroys all traces of viral nucleic acid [35]. More precise studies of PB19 inactivation by different treatment regimens await infectivity studies with developed in vitro culture systems for PB19 [36, 37].

Laboratory Screening of Blood and Blood Products

The detection of positive blood or blood products for PB19 usually depends on the method used. The greater sensitivity is for the polymerase chain reaction (PCR) . Blood units containing less than 10^6 copies, equivalent to approximately 5 pg of viral antigen would be unlikely to have been detected by the most sensitive enzyme-linked immunosorbent assay-based methods (ELISA). Hybridization methods typically detect 0.1 to 1 pg of B19 DNA (corresponding to approximately 2 x 10^4 to 2 x 10^5 copies). A hybridization assay with a chemiluminescent probe showed a sensitivity of 20 fg of B19 DNA, or 4,000 copies[38], but even this method would not have been suitable for testing large pools because it is considerably less sensitive than the PCR (5 copies per ml of plasma). Assays that are based on exploitation of the P antigen receptor of B19, known as receptor-mediated haemagglutination, have been proposed as a cheap way to screen plasma and to apparently detect whole virus; however, assay sensitivity is quite low, especially when compared to that of PCR [39-41]. A novel immunoassay that is capable of detecting 1–2 x 10^6 genome equivalents of PB19 antigen (whole virus), in the presence or absence of PB19 IgG/M, is currently under development [42] and should find an application in either mini-pool or diagnostic screening, as an adjunct to PCR screening.

Control of Viral in Blood Products

Blood and plasma units may be collected from a selected donor population, contributing to the exclusion of individuals at risk of carrying infectious agents. Collection and testing procedures of blood and plasma that are designed to exclude donations contaminated by viruses provide a solid foundation for the safety of all blood products. Each blood/plasma unit is individually screened to exclude donations positive for a direct (e.g., viral antigen) or an indirect (e.g. anti-viral antibodies) viral marker. As infectious donations, if collected from donors in the testing window period, can still be introduced into manufacturing plasma pools. So, the production of pooled plasma products requires a specific approach that integrates additional viral reduction procedures. Prior to the large-pool processing, samples of each donation for fractionation are pooled ('mini-pool') and subjected to a nucleic acid amplification test (NAT) by, for example, the polymerase chain reaction (PCR) to detect viral genomes (in Europe: HCV RNA plasma pool testing is now mandatory).

Any individual donation found PCR positive is discarded before the industrial pooling. The pool of eligible plasma donations (which may be 2000 liters or more) may be subjected to additional viral screening tests, and then undergoes a series of processing and purification steps that, for each product, comprise one or several reduction treatments to exclude HIV, HBV HCV and other viruses.

Viral inactivation treatments most commonly used are solvent-detergent incubation and heat treatment in liquid phase (pasteurization). Nanofiltration (viral elimination by filtration), as well as specific forms of dry-heat treatments, have gained interest as additional viral reduction steps coupled with established methods. Viral reduction steps have specific advantages and limits that should be carefully balanced with the risks of loss of protein activity and enhancement of epitope immunogenicity. Due to the combination of these overlapping strategies, viral transmission events of HIV, HBV, and HCV by plasma products have become very rare. Nevertheless, the vulnerability of the plasma supply to new infectious agents requires continuous vigilance so that rational and appropriate scientific promptly counter measures against emerging infectious risks can be implemented [43, 44].

Modern virucidal methods, such as cross-linked starch-iodine, dry super-heating for a long period, 80°C for 72 hours, 76°C or 90°C for 10 hours, [45] reduce by more than 7 logs the non enveloped virus load in spiked plasma pools. Nanofiltration seems to be very effective in reducing by 5.7 to 7.8 logs the bovine parvovirus load [46]. Nevertheless, PB19 DNA has been detected in commercial clotting factor concentrates, either untreated or submitted to various kinds of chemical and physical virucidal procedures.

Two studies carried out in Florence at different times, in 1991and in 1997[47], on untreated clotting factor concentrates and on concentrates submitted to different virucidal procedures , have shown that PB19 DNA was present in 37% of untreated and in 22% of treated products, especially in those treated with the solvent detergent technique.

A recently described iatrogenic transmission of Parvovirus B19 is a thread for contaminating blood derivative. It is a frequent contaminant of human blood and plasma derivatives and iatrogenic transmission of PB19 infection has been shown to occur through the administration of contaminated products. Manufacturing procedures, generally used for removal or inactivation of enveloped viruses (HIV, HCV and HBV) are not always effective in the elimination of PB19 virus. The risk of PB19 transmission by blood products is enhanced by a high virus titer in the infected donor, by pooling of a large number of donations, and by the virus's resistance to effective inactivation methods such as heat and solvent-detergent treatments. Solvent-detergent treatments fail to destroy PB19 capsids because of the absence of a lipid-envelope, and heat treatments (pasteurization and dry-heat methods) cannot guarantee a complete viral inactivation because of the variable heat sensitivity of the virus.

There is data about the ability to inactivate PB19 by liquid-heat treatment at 60 degrees C for 10 hours that was incorporated in the manufacturing process of intravenous human immunoglobulin preparations. The results showed that PB19 was rapidly inactivated under the conditions used for the liquid-heat treatment [48]. Another data determined that nanofilteration is effective for removal of PB19. It assumed that, since antibodies against these viruses, HAV and B19, are constantly present in large plasma pools, it is conceivable that antibody binding will lead to an increase in size of the viruses due to a corona of antibodies on the virus surfaces. These data indicates that viruses with bound antibodies are efficiently eliminated by nanofiltration by use of filters having nominal pore sizes larger than the diameter of the respective free virions [49].

Another control measure to be acknowledged that many manufacturers now perform PB19 PCR on plasma mini-pools, in order to eliminate high B19 viral load plasma [50]. Nucleic acid amplification techniques (NAT) are the methods of choice for the detection of viruses, due to their high specificity and sensitivity. NAT assays are beneficial tools for the identification of contaminated mini-pools or plasma pools and the quantification of PB19 contamination. They may also be valuable for testing the removal of PB19 virus during manufacturing: since the virus may not be completely inactivated or removed by chemical or physical treatments, residual PB19 contamination should always be checked. PCR screening of blood products has been shown to facilitate removal of PB19 PCR-positive donations from a plasma pool of 6000, resulting in a 10–100-fold decrease in viral load [10]. However, the level of PB19 DNA does not necessarily correlate to the rate of infectivity, as has been demonstrated previously for the canine parvovirus; in this case, heat treatment was found to reduce canine parvovirus infectivity by >100-fold, despite the fact that the PCR assay titer was unaltered [50, 51].

Nonetheless, the issue of whether high-risk populations, such as pregnant women, immunocompromised patients and people with chronic anaemia, should undergo administration of any PB19-containing products while the level of infectious PB19 DNA is unknown and mini-pool screening is not mandatory must be addressed. The aforementioned availability of an International Standard preparation of PB19 DNA [52], in addition to a number of compatible and quantitative PB19 PCR detection systems [9,26,50], should alleviate problems caused by ambiguity between results from laboratories that use various methods of measuring and expressing PB19 DNA levels and help to determine the infectious dose for PB19.

It is highly significant that, in the Netherlands, a recommendation has been made that individual donor screening for PB19 IgG to identify individual donors with continually high antibody levels (at $t = 0$ and 6 months) should be initiated [53,54]. Selected individuals who maintain high B19 IgG levels will subsequently form a panel of plasma or blood-product donors for high-risk recipients, such as immunocompromised individuals, thereby minimizing the risk of PB19 transmission from acutely infected, although asymptomatic, donors. It is our view that the introduction of such a screening algorithm sets the standard for blood-product safety in the future, specifically with respect to minimizing the risk of PB19 transmission.

Conversely, as a large number of blood donations make up the plasma pools used to produce plasma derivatives, clotting-factor concentrates may be contaminated. Studies have detected PB19 in two of three unheated batches of factor preparations and in 20 to 25% of solvent or detergent-treated batches, while the fractionation process used to obtain albumin preparations was apparently more efficient at eliminating virus [55], but PB19 was still found in 3 of 12 batches in one study [56]. Recent study has been made to evaluate the ability to inactivate PB19 of liquid-heat treatment at 60 degrees C for 10 hours that was incorporated in the manufacturing process of intravenous human immunoglobulin preparations; showed that PB19 was rapidly inactivated under these conditions [57].

Even after the introduction of virus-inactivated clotting-factor concentrates, a PB19 seroprevalence among hemophiliacs of 90% has been observed, with correlation to the amount of clotting factor received [58]. PB19 may also infrequently be transmitted by bone marrow (BM) [59] and blood-derived products such as platelets [60] intravenous immunoglobulin [61], and fibrin products[62], and PB19 infection and seroconversion have been observed in patients after receiving solvent or detergent-treated plasma units [63].

Conclusion

Notwithstanding the good progress attained in the attempt to attenuate the infectivity of enveloped viruses, the risk of transmission of parvovirus B 19 is still quite high, 40%, according to the results of the most recent prospective study . Because of its resistance to chemical or physical virucidal methods, parvovirus B19 represents an important marker of plasma pool contamination by viruses. This issue needs more extended studies.

References

[1] Allain JP, Stramer SL, Carneiro-Proietti AB, Martins ML, Lopes da Silva SN, Ribeiro M, Proietti FA, Reesink HW. Transfusion-transmitted infectious diseases. *Biologicals.* 2009, 37(2):71-7.

[2] Cohen, B. J., A. M. Field, S. Gudnadottir, S. Beard, and J. A. Barbara. Blood donor screening for parvovirus B19. *J. Virol. Methods.* 1990, 30:233-238.

[3] Cunningham, D. A., J. R. Pattison, and R. K. Craig. Detection of parvovirus DNA in human serum using biotinylated RNA hybridisation probes. *J. Virol. Methods.* 1988, 19:279-288.

[4] Mori, J., A. M. Field, J. P. Clewley, and B. J. Cohen. Dot blot hybridization assay of B19 virus DNA in clinical specimens. *J. Clin. Microbiol.* 1989, 27:459-464.

[5] Musiani, M., M. Zerbini, D. Gibellini, G. Gentilomi, S. Venturoli, G. Gallinelia, E. Fern, and S. Girotti. Chemiluminescence dot blot hybridization assay for detection of B19 parvovirus DNA in human sera. *J. Clin. Microbiol.* 1991, 29:2047-2050.

[6] Salimans M M, Holsappel S, van de Rujke F M, Jiwa N M, Raap A K, Weiland H T. Rapid detection of human parvovirus B19 DNA by dot-hybridization and the polymerase chain reaction. *J. Virol. Methods.* 1989, 23: 19-28.

[7] Clewley J P. Polymerase chain reaction assay of parvovirus B19 DNA in clinical specimens. *J. Clin. Microbiol.* 1989, 27: 2647-2651. 9.

[8] Koch W C, and Adler S P. Detection of human parvovirus B19 DNA by using the polymerase chain reaction. *J. Clin. Microbiol.* 1990, 28:65-69.

[9] Thomas I, Di Giambattista M, Gérard C, Mathys E, Hougardy V, Latour, B. Branckaert T, Laub R. Prevalence of human erythrovirus B19 DNA in healthy Belgian blood donors and correlation with specific antibodies against structural and non-structural viral proteins. *Vox Sang.* 2003, 84, 300–307

[10] Prowse C, Ludlam C A, Yap PL. Human parvovirus B19 and blood products. *Vox Sang.* 1997, 72, 1–10

[11] O'Neill H J & Coyle P V. Two anti-parvovirus B 19 IgM capture assays incorporating a mouse monoclonal antibody specific for B 19 viral capsid proteins VP 1 and VP 2. *Arch. Virol.* 1992,123, 125–134.

[12] Davenport R, Geohas G, Cohen S, Beach K, Lazo A, Lucchesi K, Pehta J. Phase IV study of Plas+®SD: hepatitis A (HAV) and parvovirus B19 (B19) safety results. *Blood.* 2000, 96: 1942- 1947.

[13] Eis-Hübinger A M, Oldenburg J, Brackmann H H, Matz B, Schneweis K E. The prevalence of antibody to parvovirus B19 in hemophiliacs and in the general population. *Zentbl. Bakteriol.* 1996, 284: 232–240.

[14] Blümel J, Schmidt I, Willkommen H, Löwer J. Inactivation of parvovirus B19 during pasteurization of human serum albumin. *Transfusion*. 2002, 42: 1011–1018.

[15] Zanella A, Rossi F, Cesana C, et al. Transfusion transmitted human parvovirus B19 infection in a thalassemic patient. *Transfusion*. 1955, 35:769-772.

[16] Azzi A, Zakrzewska K, Bertoni E, et al. Persistent parvovirus B19 infections with different clinical outcome in renal transplant recipients: diagnostic relevance of PCR and of quantification of B19 DNA in sera. *Clin. Microbiol. Infect.* 1996, 2:105-108.

[17] Jordan J, Tiangco B, Kiss J, et al: Human parvovirus B 19: Prevalence of viral DNA in volunteer blood donors and clinical outcomes of transfusion recipients. *Vox Sang*. 1998, 75: 97-102.

[18] McOmish F, Yap PL, Jordan A, et al: Detection of parvovirus B 19 in donated blood: A model system for screening by polymerase chain reaction. *J. Clin. Microbiol.* 1993, 31:323-328,

[19] Kurtzman G J, Cohen B J, Field A M, Oseas R, Blaese R M, Young N S. Immune response to B19 parvovirus and an antibody defect in persistent viral infection. *J. Clin. Invest.* 1989, 84:1114-1123.

[20] Mortimer P P Transmission of serum parvovirus-like virus by clotting-factor concentrates. *Lancet*. 1983, ii:482-484.

[21] Rollag H, Patou G, Pattison J R, Degre M, Evensen S A, Froland SS, Glomstein A. Prevalence of antibodies against parvovirus B19 in Norwegians with congenital coagulation factor defects treated with plasma products from small donor pools. *Scand. J. Infect. Dis.* 1991, 23:675-679.

[22] Williams M D, Cohen B J, Beddall A C, Pasi K J, Mortimer P P, Hill F G. Transmission of human parvovirus B19 by coagulation factor concentrates. *Vox Sang*. 1990, 58: 177-181.

[23] Schwarz T F, Roggendorf M, Hottentrager B, Stolz W, Schwinn H. Removal of parvovirus B19 from contaminated factor VIII during fractionation. *J. Med. Virol.* 1991, 35:28-31.

[24] Williams MD, Cohen BJ, Beddall AC et al. Transmission of human parvovirus B19 by coagulation factor concentrates. *Vox Sanguis*. 1990; 58: 177-181.

[25] Santagostino E, Mannucci P M, Gringeri A, Azzi A & Morfini M. Eliminating parvovirus B19 from blood products. *Lancet.*1994, 343: 798. 798.

[26] Daly P, Corcoran A, Mahon B. P., Doyle S. High-Sensitivity PCR Detection of Parvovirus B19 in Plasma. *J. of Clin. Microbiol.* 2002, 40: 6/ 1958-1962, .

[27] Kleinman SH, Glynn SA, Lee TH, Tobler LH, Schlumpf KS, Todd DS, Qiao H, Yu MY, Busch MP. A linked donor-recipient study to evaluate parvovirus B19 transmission by blood component transfusion. *Blood.* 2009, 225706v1

[28] Saldanha J & Minor P. Detection of human parvovirus B19 DNA in plasma pools and blood products derived from these pools: implications for decency and consistency of removal of B19 DNA during manufacture. *Br. J. of Haematol.* 1996; 93: 714±719.

[29] Prowse C, Ludlam C A and Yap P L. Human Parvovirus B19 and Blood Products. *Vox Sanguinis.* 2003, 72: 1/ 1 – 10.

[30] Anderson MJ, Jones SE, Fisher-Hoch SP et al. Human parvovirus, the cause of erythema infectiosum (fifth disease)? . *Lancet.* 1983; I: 1378.

[31] Serjeant GR, Topley JM, Mason K et al. Outbreak of aplastic crisis in sickle cell anaemia associated with parvovirus-like agent. *Lancet.* 1981; ii: 595-597.

[32] Morfini M, Longo G, Rossi Ferrini P, Azzi A, Zakrewska C, Ciappi S, Kolumban P. Hypoplastic anemia in a hemophiliac first infused with a solvent/detergent treated factor VIII concentrate: the role of human B19 parvovirus. *Am. J. Hematol.* 1992, 39:149-150.

[33] Pattison JR, Jones SE, Hodgson J et al. Parvovirus infections and hypoplastic crisis in sickle-cell anaemia. *Lancet.* 1981; i: 664-665.

[34] Yoto Y, Kudoh T, Haseyama K, Suzuki N, Oda T, Katoh T, Takahashi T, Sekiguchi S & Chiba S. Incidence of human parvovirus B19 DNA detection in blood donors. *Br. J. Haematol.* 1995, 91, 1017–1018.

[35] Salimans M M, Holsappel S, van de RUjke F M, Jiwa N M, Raap A K, Weiland H T. Rapid detection of human parvovirus B19 DNA by dot-hybridization and the polymerase chain reaction. *J. Virol. Methods.* 1989, 23:19-28.

[36] Cossart YE, Field AM, Cant B & Widdows D. Parvovirus-like particles in human sera. *Lancet.* 1975; i: 72-73.

[37] Anderson MJ, Lewis E, Kidd IM et al. An outbreak of erythema infectiosum associated with human parvovirus infection. *Journal of Hygiene.* 1984; 93: 85-93.

[38] Bonvicini F, Gallinella G, Gentilomi GA, Ambretti S, Musiani M, Zerbini M. Prevention of iatrogenic transmission of B19 infection: different approaches to detect remove or inactivate virus contamination. *Clin. Lab.* 2006; 52(5-6):263-8.

[39] Cohen B J & Bates C M. Evaluation of 4 commercial test kits for parvovirus B19-specific IgM. *J. Virol. Methods.* 1995, 55, 11–25.

[40] Sato H, Takakura F, Kojima E, Fukada K, Okochi K. & Maeda Y. Screening of blood donors for human parvovirus B19. *Lancet.* 1995, 346, 1237–1238

[41] Wakamatsu, C., Takakura, F., Kojima, E. et a. Screening of blood donors for human parvovirus B19 and characterization of the results. *Vox Sang.* 1999, 76, 14–21

[42] O'Keeffe, S., O'Leary, D., Doyle, S., Kilty, C. & Kerr, S. The detection of parvovirus B19 in human sera using antigen-capture EIA. Poster presented at the *Meeting of the Society for General Microbiology (Irish Branch)*, 2003. National University of Ireland, Maynooth, Co. Kildare, Ireland, 24–25 April 2003.

[43] Burnouf-Raldosevich M, Appourchaux R Huart JJ, et al: Nanofiltration, a new specific virus elimination method applied to high-purity factor IX and factor XI concentrates. *Vox Sang.* 1994, 67:132-138

[44] Burnouf T, Radosevich M. Reducing the risk of infection from plasma products: specific preventative strategies. *Blood Rev.* 2000; 14:2/94-110.

[45] Hart HF, Hart WG, Crosstey J, et al: Effect of terminal (dry) heat treatment on non-enveloped viruses in coagulation factor concentrates. *Vox Sang.* 1994, 67: 345-350.

[46] Lefrere JJ, Mariotti M, Thauvin M: B 19 parvovirus DNA in solvent detergent treated antihemophilia concentrates. *Lancet.* 1994, 343:211-212

[47] Eis-Hubinger AM, Aberham C, Pendl, C, Gross P, Zerlauth G & Gessner M. A quantitative, internally controlled real-time PCR assay for the detection of parvovirus B19 DNA. *J. Virol. Methods.* 2001, 92, 183–191.

[48] Zakrzewska K, Azzi A, Patou G, et al: Human parvovirus B 19 in clotting factor concentrates: B 19 DNA detection by the nested polymerase chain reaction. *Br. J. Haematol.* 1992, 81:407-412

[49] Yunoki M, Urayama T, Tsujikawa M, et al. Inactivation of parvovirus B19 by liquid heating incorporated in the manufacturing process of human intravenous immunoglobulin preparations. *Br. J. Haematol.* 2005, 128:401-4.

[50] Omar A, Kempf C. Removal of neutralized model parvoviruses and enteroviruses in human IgG solutions by nanofiltration. *Transfusion.* 2002; 42,8:1005-10.

[51] Aberham C, Pendl C, Gross P, Zerlauth G & Gessner M. A quantitative, internally controlled real-time PCR assay for the detection of parvovirus B19 DNA. *J. Virol. Methods*. 2001, 92, 183–191

[52] Hart H F, Hart W G, Crossley J, Perrie A M, Wood D J, John A. & McOmish F. Effect of terminal (dry) heat treatment on non-enveloped viruses in coagulation factor concentrates. *Vox Sang.* 1994, 67, 345–350

[53] Saldanha J, Lelie N, Yu M W, Heath A & B19 Collaborative Study Group. Establishment of the first World Health Organization International Standard for human parvovirus B19 DNA nucleic acid amplification techniques. *Vox Sang.* 2002, 82, 24–31.

[54] Health Council for the Netherlands. *Blood Products and Parvovirus B19: 'Alerting' Advisory Report.*2002, (publication no. 2002/07; ISBN 90-5549-432-1). The Hague: Health Council for the Netherlands.

[55] Groeneveld, K. & van der Noordaa, J. Blood products and parvovirus B19. *Neth. J. Med.* 2003, 61, 154–156

[56] Yee T T, Cohen B J, Pasiasi K J, Lee C A. Transmission of symptomatic parvovirus B19 infection by clotting factor concentrate *Br. J. of Haematol.* 2003, 93: 2/ 457 - 459

[57] Rollag H, Solheim BG and Svennevig JL. Viral safety of blood derivatives by immune neutralization. *Vox Sang.* 1998 , 74:1/ 213-17.

[58] Yunoki M, Urayama T, Tsujikawa M, Sasaki Y, Abe S, Takechi K, Ikuta K.Inactivation of parvovirus B19 by liquid heating incorporated in the manufacturing process of human intravenous immunoglobulin preparations. *Br. J. Haematol.* 2005 ; 128:3/401-4

[59] Rollag H, Patou G, Pattison JR, et al. Prevalence of antibodies against parvovirus B19 in Norwegians with congenital coagulation factor defects treated with plasma products from small donor pools. *Scand. J. Infect. Dis.* 1991.23:675-79.

[60] Heegaard ED and Laub PB. Parvovirus B19 transmitted by bone marrow. *Br. J. Haematol.* 2000, 111:659-61.

[61] *Cohen BJ, Beard S, Knowles WA, Ellis JS, Joske D, Goldman JM, Hewitt P, Ward KN.* Chronic anemia due to parvovirus B19 infection in a bone marrow transplant patient after platelet transfusion. *Transfusion. 1997, 37: 9/947-52.*

[62] Erdman DD, Anderson BC, Torok TJ, et al. Possible transmission of parvovirus B19 from intravenous immune globulin. *J. Med. Virol.* 1997, 53: 233-36.

[63] Hino M, Ishiko O, Honda KI, et al. Transmission of symptomatic parvovirus B19 infection by fibrin sealant used during surgery. Br. J. Haematol. 2000,108:194-95.

[64] Brown KE, Young NS, Alving BM, et al. Parvovirus B19: implications for transfusion medicine. Summary of a workshop *Transfusion.* 2001, 41: 130-35.

Chapter 6

Occupational Infection by PB19

Abstract

PB19 is a common virus that is spread worldwide, and the seroprevalence increases with age, so that up to 15% of preschool children, 50% of younger adults and about 85% of the elderly show serologic evidence of past infection. In developing countries the seroprevalence has been shown to be a little higher, probably because of poor and crowded living standards, whereas in isolated communities seroprevalence figures are below 10%. Infection appears to confer lifelong immunity to immunocompetent hosts by the presence of neutralizing antibodies. There is seasonal variation of the incidence of infection according to climates, being more common during winter and early spring. Epidemics are reported at intervals of about 3–4 years, especially for outbreaks of erythema infectiosum and PB19-related disease. PB19 is normally transmitted through the respiratory route As most infections occur in children aged 5–15 years, adults in direct contact with children are at risk like parents of children in that age group, or those working at day care centers or schools

Parvovirus causes a wide spectrum of clinical complications ranging from mild, self-limiting erythema infectiosum in immunocompetent children to lethal cytopenias in immunocompromised patients and intrauterine foetal death in primary infected pregnant women. The exposure of individuals lacking neutralizing antibodies to infected individuals can lead to subsequent symptomatic infections. Whether to consider PB19 as one of occupational pathogen especially for medical workers and school workers is controversy.

Keywords: PB19, occupational hazard, school workers, medical care workers.

Introduction

Human parvovirus B19 (PB19) infection, early diagnosed as fifth disease (erythema infectiosum: EI) mainly occurs in children. Though the prevalence of anti-PB19 antibodies in adults aged 20 to 40 years range from 30 to 40%. Susceptible adults are occasionally infected through contact with a child harboring acute PB19 infection, mean while the risk is considered low. In particular, nosocomial infection in nursing staff members acquired through contact with a patient with haemolytic anemia has been reported. There is more risk in school employee. The data concerning complications of such condition is still insufficient. Nevertheless, control of infection is required with good hygiene measures.

Risk of PB19 Infection in Health Care Workers

Previous data has shown that hospital workers may not at high risk of infection with PB19 during endemic periods [1]. However, other data reported that hospital acquisition and transmission have been reported after exposure to patients with acute infection [2-5].

Other source of infection to health care workers is transmission of PB19 to health care workers from their children and then infection is further transmitted to others working in the same ward. Moreover, this lead to wide spread of PB19 in health care workers. Hospital office staff members rarely work in a nursing ward, but they may have direct contact with nurses in a staff room. Therefore, transmission of the virus probably occurred in a staff room through person-to-person contact or via contaminated materials.

Nevertheless, most reported outbreaks have been limited to a specific sector of a hospital [3-5] like maternity ward [6]. It is in fact that a generalized outbreak throughout the community and hospital occur with similar infection rates. In an outbreak in a pediatric hospital, 12 of 32 susceptible health care workers (38%) developed disease [7]. Rates of infection were highest among nurses who were exposed to non isolated patients soon after admission and exceeded those reported from family-based outbreaks [8] and outbreaks in day care centers [9,10]. Outbreak-related incidence rates ranging from 27% to 47% have been described in an adult ward [11], a developmental center [12], and a neurosurgical intensive care unit [13]. A normal child was born to an employee who became infected while pregnant [14]. One study found that no transmission had occurred, possibly because the source case was a chronic secretor with a relatively low viral load [15]; another suggested that community-based transmission had a significant role [16].

The fetal risk of non immune hydrops after maternal PB19 infection must be very low. As a consequence, exclusion of pregnant women from the workplace during endemic periods with seasonal clusters of cases is not justified. Weekly fetal ultrasound evaluation in these cases carries a low yield.

Data has reported a PB19 outbreak at a general hospital (Hospital Central Dr Ignacio Morones Prieto, San Luis Potosí, Mexico). At same time several medical students attending the hospital reported symptoms suggestive of PB19 infection [17]. The acute infection rate among susceptible students that was found (36.1 %) is similar to that reported in hospital outbreaks by others. It was reported a conversion rate 44 % among susceptible hospital workers in close contact with patients with sickle-cell disease and acute PB19-associated aplastic crisis [18]. In that report, only close contacts to the affected patients were evaluated. Thirty-three per cent of susceptible subjects studied during another PB19 outbreak at a children's ward were found to have acute infection [4, 6]. In same period the infection rates reported were 23-30 % among susceptible subjects inside and outside of the hospital during a community-wide PB19 outbreak. Once again there is confusion whether these infections occur in hospital or outside.

In one study medical students attending surgical ward have significantly higher rates of acute infection (62.5 %) among susceptible subjects compared to students attending other wards, or not attending the hospital. This high rate of acute infection among students compares to the previous infection rate 67 % reported in susceptible household contacts of PB19 IgM-positive erythema infectiosum cases [19]. Of interest, the first subjects to be identified with PB19-associated symptoms were anesthesiology residents, and the subject considered to be the index case for the hospital outbreak was an intern rotating on the surgical service. There is no obvious data to identify factors that accounted for the higher infection rate in students attending the surgical wards. Whether closer contact among these students or exposure to a patient with prolonged viral excretion, such as an immunocompromised patient, accounted for this high infection rate is not determined [17].

Infection with PB19 has special dangerous aspects for pregnant women. Moreover, previous reports have shown that women have a higher infection rate [20]. It has been suggested that women are more likely to be exposed to children from whom they may acquire the infection. In published data, it was found that pregnant women with more children have higher seroprevalence rate [21]. It should be pointed out that we were only provided with information regarding the number of children at home and not the extent of contact with them, which may be a more reliable marker for the risk of acquiring an infection. Women, especially mothers, are more likely to have closer contact with children and, thus, may be at higher risk of acquiring B19 infection. The higher frequency of acute infection in women is of importance, since pregnant women are at risk of delivering a newborn with hydrops or even have other adverse events during pregnancy [20]. A trial to define risk factors about patient's source of PB19 is controversy. It has been described a PB19 outbreak among hospital workers without known exposure to patients with erythema infectiosum or aplastic crisis. These findings suggest that transmission among hospital staff may be responsible for some PB19 outbreaks among health-care workers [23].

Risk of PB19 Infection in School Employee

Infection with PB19 is common during childhood and school personnel are at risk of acquiring this infection [24].

In non outbreak settings, the risk of school employees is 13 times higher than that of hospital workers [25]. Workers having contact with younger children and with larger numbers of children had higher rates of infection. In general, high rates of infection among teachers indicate that exposure to this illness occurs in these work environments. So, PB19 infection is an occupational risk for day-care workers and school personnel.

Infection control measures for Occupational Infection with PB19

1. Hospital Care Workers

Several infection control measures are put into place at the hospital once the outbreak is identified,. All cases fulfilling a clinical case definition for acute B19 infection should be identified and sera obtained for confirmation of acute infection. Droplet isolation is advisable for seven days after the onset illness in hospitalized patients. However, there is a risk of nosocomial transmission from patients with transient aplastic crises (TAC) and from immunodeficient patients with chronic PB19 infection. These patients should be considered infectious and placed on isolation precautions for the duration of their illness or until the infection has been cleared.

Information regarding the outbreak should be provided to all services. The number of new cases should be assessed weekly until no new cases are identified.

Infection control measures, in particular hand-washing, are reinforced. All hospital staff is encouraged to wear masks.

Screening of HCWs to identify those who are susceptible to infection is not justified in general, but may be important in certain specific circumstances. Specifically, laboratory workers who are to work with infectious materials known to contain parvovirus B19 virus should be screened to determine whether they are susceptible. HCWs that work with at-risk patients and have significant contact with a confirmed case should have their serostatus determined.

A HCW who is seronegative should be excluded from further contact with at-risk patients until either the rash appears or until 15 days from the last date of significant contact with the case. The diagnosis can be confirmed or excluded serologically 21 days from the last contact. Serology can be performed when the rash appears or, if there is no clinical illness, a subclinical infection should be excluded by testing for IgG and IgM antibodies 21 days after the last contact.

If parvovirus B19 infection is confirmed in a HCW then the implications need to be considered for patients at-risk from parvovirus infection who were in contact with that HCW during the 7 days before onset of the rash.

HCW should not be caring for patients when they may have an infectious disease indicated by influenza-like symptoms, a fever or rash. The exclusion of a symptomatic parvovirus B19-infected HCW may offer small practical benefit since the peak infective period will have passed by the time the rash and associated symptoms appear. However, in most cases serological confirmation of parvovirus B19 infection will not be immediately available, and the difficulty of making an accurate diagnosis on clinical grounds means that it will not be possible to differentiate between parvovirus and other illnesses which can cause nosocomial outbreaks such as rubella or measles. Therefore, a HCW who may have a parvovirus infection should be advised to stay off work until they no longer present a potential risk to patients or colleagues.

Overall, the risks of a pregnant woman becoming infected are greater outside the healthcare setting than within it, particularly if she has children or works with children. Women who were in contact with the HCW outside the infectious period or greater than 20 weeks pregnant can be reassured.

Women, who are less than 20 weeks pregnant, may have their susceptibility determined by testing serum for antibodies to parvovirus B19. The consent of any women whose sera are to be tested should be sought before testing. About 60% of pregnant women will be immune to parvovirus B19 because of previous infection and can be reassured that they are at no risk.

The estimated ratio 40% of women who are susceptible will not necessarily become infected after contact with another infected person as this depends on the nature of the contact. They will require serological follow-up to identify who has become infected and will then require specialist referral. Nevertheless, it is avisable that susceptible pregnant provided with paid leave until the outbreak is ceased.

There is a theoretical risk of immunocompromised patients becoming infected through contact with an infected HCW. Some specialised units may consider staff screening to identify staff who are seropositive and hence able to care for infectious patients without presenting an infection control hazard.

This might avoid having to treat each case of contact between HCW and someone with a rash as an incident of potential transmission, if this is causing excessive "fire-fighting" activity in a particular unit.

2. School Employee

If a case has been in contact with a woman in the first 20 weeks of pregnancy or an immunocompromised individual, further follow up may be required and the Health Protection Unit should be contacted to advice. This is particularly important where the case is a healthcare worker who may have had exposure to vulnerable groups.

School exposure: This advice applies equally to employees working in settings such as primary schools where the rates of parvovirus infection may be higher than in other settings. If there is an outbreak in a school, employees such as teachers who are in contact with affected children and who are less than 20 weeks pregnant should also seek medical advice.

Only during an outbreak should exclusion of employees who are less than 20 weeks pregnant and in close contact with children be considered. The employee should first find out whether or not she is susceptible to parvovirus PB19. The employee should be informed that the outbreak probably reflects the situation in the community at large and that avoiding contact with children at school will not necessarily reduce the risk of infection (Coated from Louisiana Office of Public Health – Infectious Disease Epidemiology Section-Infectious Disease Control Manual

Conclusion

Human parvovirus B19 (PB19) infections are common in children, and maternal infection of pregnant women may cause fetal death, risk factors for PB19 infections for hospital and school employees were identified during an endemic period and in presence of sporadic infections. Hospital outbreak may either reflect transmission outside the hospital or acquisition of infection inside hospital. Contaminated environmental surfaces and patients with acute infections are identified as one potential source for transmission of parvovirus B19. Those in daily contact with school-age children had a higher risk of increased annual occupational risk for PB19 infection. Several infection control measures are put into place when an outbreak arises. Infection control depends upon site of outbreak and the condition of the exposed individuals.

References

[1] Adler S P, Manganello A M, Koch W C, Hempfling S H. & Best A M. Risk of human parvovirus B19 infections among school and hospital employees during endemic periods. *J. Infect. Dis.* 1993, 168, 361–368

[2] Evans J P M, Rossiter M A, Kumaran T O, Marsh G W & Mortimer P P. Human parvovirus aplasia: case due to cross infection in a ward. *Br. Med. J. (Clin. Res. Ed)*. 1984, 288, 681. 681.

[3] Bell L M, Naides S J, Stoffman P, Hodinka R L & Plotkin S A. Human parvovirus B19 infection among hospital staff members after contact with infected patients. *N. Engl. J. Med.* 1989, 321, 485–491

[4] Pillay D, Patou G, Hurt S, Kibbler C C & Griffiths P D. Parvovirus B19 outbreak in a children's ward. *Lancet*. 1992, 339, 107–109.

[5] Farr R W, Hutzel D, D'Aurora R & Rugino T. Parvovirus B19 outbreak in a rehabilitation hospital. *Arch. Phys. Med. Rehabil.* 1996,77, 208 210.

[6] Dowell S F, Török T J, Thorp J A, Hedrick J, Erdman DD, Zaki S R, Hinkle C J, Bayer W L & Anderson L J. Parvovirus B19 infection in hospital workers: community or hospital acquisition? *J. Infect. Dis.* 1995, 172, 1076–1079

[7] Bell LM, Naides SJ, Stoffman P, Hodinka RL, Plotkin SA. Human parvovirus B19 infection among hospital staff members after contact with infected patients. *N. Engl. J. Med.* 1989; 321:485-91.

[8] Cartter ML, Farley TA, Rosengren S, Quinn DL, Gillespie SM, Gary GW, et al. Occupational risk factors for infection with parvovirus B19 among pregnant women. *J. Infect. Dis.* 1991; 163:282-5.

[9] Gillespie SM, Cartter ML, Asch S, Rokos JB, Gary GW, Tsou CJ, et al. Occupational risk of human parvovirus B19 infection for school and day-care personnel during an outbreak of erythema infectiosum. *JAMA.* 1990; 263:2061-5.

[10] Pickering LK, Reves RR. Occupational risks for child-care providers and teachers [Editorial]. *JAMA.* 1990; 263:2096-7.

[11] Seng C, Watkins P, Morse D, Barrett SP, Zambon M, Andrews N, et al. Parvovirus B19 outbreak on an adult ward. *Epidemiol. Infect.* 1994; 113:345-53.

[12] Lohiya GS, Stewart K, Perot K, Widman R. Parvovirus B19 outbreak in a developmental center. *Am. J. Infect. Control.* 1995; 23:373-6.

[13] Shishlba T, Matsunaga Y. An outbreak of erythema infectiosum among hospital staff members including a patient with pleural fluid and pericardial effusion. *J. Am. Acad. Dermatol.* 1993; 29:265-7.

[14] Harrison J, Jones CE. Human parvovirus B19 infection in healthcare workers. *Occup. Med.* (Oxf). 1995; 45:93-6.

[15] Koziol DE, Kurtzman G, Ayub J, Young NS, Henderson DK. Nosocomial human parvovirus B19 infection: lack of transmission from a chronically infected patient to hospital staff. *Infect. Control Hosp. Epidemiol.* 1992; 13:343-8.

[16] Dowell SF, Torok TJ, Thorp JA, Hedrick J, Erdman DD, Zaki SR, et al. Parvovirus B19 infection in hospital workers: community or hospital acquisition? *J. Infect. Dis.* 1995; 172:1076-9.

[17] Noyola D E, Padilla-Ruiz M L, Obregón-Ramos M G, P Zayas and B Pérez-Romano. Parvovirus B19 infection in medical students during a hospital outbreak . *Med. Microbiol.* 53. 2004, 141-146.

[18] Bell, L. M., Naides, S. J., Stoffman, P., Hodinka, R. L. & Plotkin, S. A. Human parvovirus B19 infection among hospital staff members after contact with infected patients. *N. Engl. J. Med.* 1989, 321, 485–491

[19] Chorba T, Coccia P, Holman R C et al. The role of parvovirus B19 in aplastic crisis and erythema infectiosum (fifth disease). *J. Infect. Dis.* 1986, 154, 383–393

[20] Morgan-Capner P, Wright J, Longley J P & Anderson M J. Sex ratio in outbreaks of parvovirus B19 infection. *Lancet*. 1987, ii, 98. 98.

[21] El-Sayed Z M; Hossam G · Relevance of Parvovirus B19, Herpes Simplex Virus 2, and Cytomegalovirus Virologic Markers in Maternal Serum for Diagnosis of Unexplained Recurrent Abortions. *Arch. Pathol. Lab. Med.* 2007 ;131/6:956-60.

[22] Anand A, Gray E S, Brown T, Clewley J P & Cohen B J. Human parvovirus infection in pregnancy and hydrops fetalis. *NEJM*. 1987, 316, 183–186

[23] Miyamoto K, Ogami M, Takahashi Y, Mori T, Akimoto S, Terashita H & Terashita T. Outbreak of human parvovirus B19 in hospital workers. *J. Hosp. Infect*. 2000, 45, 238–241

[24] Gillespie S M, Cartter M L, Asch, S, Rokos J B, Gary G W, Tsou C J, Hall D B, Anderson L J & Hurwitz E S. Occupational risk of human parvovirus B19 infection for school and day-care personnel during an outbreak of erythema infectiosum. *JAMA. (J. Am. Med. Assoc)*. 1990, 263, 2061–2065.

[25] Adler SP, Manganello AM, Koch WC, Hempfling SH, Best AM. Risk of human parvovirus B19 infections among school and hospital employees during endemic periods. *J. Infect. Dis.* 1993; 168:361-8.

[26] Louisiana Office of Public Health – Infectious Disease Epidemiology Section- Infectious Disease Control Manual. www.dhh.louisiana.gov/ offices/.../ Manual/GuillainBarreManual.pdf

Chapter 7

PB19 Genetic Study, Relation to Pathogenesis

Abstract

Parvovirus B19 is a common human pathogen maintained by horizontal transmission between acutely infected individuals. However, PB19 virus can also be detected in tissues throughout the life of the host, although little is understood about the nature of such persistence. In addition, the clinical presentations of infections vary individually. The species human parvovirus PB19 (B19V) can be divided into three genotypes. In this chapter, we addressed the question as to whether infection of an individual is restricted to one genotype or if genotypes variation results in different clinical presentation.

Keywords: PB19 genotypes, sequence analysis, clinical presentation.

Introduction

Parvoviridae, is a pathogenic virus distributed worldwide in the human population [1]. PB19 infection is associated with various clinical manifestations depending on interaction between viral factors and immunological and haematological status of the host [2]. Persistent PB19 infections have been documented, but their pathogenicity remains unclear [3-5].

PB19 virus shares genetic and structural characteristics of the family Parvoviridae [6]. The viral genome is a single stranded DNA molecule, about

5.6 kb in length, composed of a unique internal region, containing all the coding sequences, flanked by two inverted terminal regions. DNA molecules of either polarity are separately encapsidated at equal frequency in 25 nm icosahedral capsids composed of 60 subunits of the capsid proteins VP1 and VP2. The VP2 protein accounts for about 95% of the capsid proteins and contains the receptor binding site [7]; the VP1 protein, containing an additional unique domain at the N-terminus with respect to VP2 protein, accounts for the remaining 5% of the capsid proteins and contains a phospholipase activity necessary for viral infectivity [8].

There are three divergent genotypes (1–3) have been identified among PB19 strains [9]. The three genotypes of PB19 virus differ in nucleotide sequence by approximately 13 to 14% [10-12]. The nucleotide divergence between the genotypes is approximately 10 % and in the promoter region more than 20 %. Additionally, genotype 3 viruses cluster into two subtypes represented by the prototype strains V9 (GenBank accession no. AX003421 [GenBank]) and D91.1 (GenBank accession no. AY083234 [GenBank]) [13].

Epidemiological data reveals that genotypes 1 and 2 are found in Europe, the United States, and other Western countries, while genotype 3 is restricted to sub-Saharan Africa and South America [10,14].The prevalence of genotype 2 or genotype 3 strains appears lower than that of genotype 1 in western countries [9,15]. Consequently, few clinical and molecular data on genotype 3 strains were available until West Africa was identified as an endemic region for B19 genotype 3 infection [3]. Little is known about the origin, past epidemiology, or evolution of B19 virus. However, the lifelong persistence of viral DNA in tissues [16-20] may reveal what variants circulated in previous decades of life when primary infections occurred [21]. Sequences of both genotype 1 and 2 were found in tissues of persistently infected individuals, although genotype 2 detection was strictly limited to those born before 1973, indicating that the circulation of this genotype ceased in Finland and Germany after this date [22]. A similar restriction of genotype 2 to those born before 1963 was found in study subjects from Scotland [23], indicating some commonality in the transmission networks and dynamics of PB19 over large areas of Europe.

Variants of parvovirus B19 are currently grouped into three genotypes: 1 (reference PB19 strains), 2 and 3. It has been speculated that certain genotype could contribute to specific pathological condition. It has been evidenced that isolate K71 of genotype 2 is more prevalent in skin than the conventional PB19 1 genotype [24].

Genotype 2 viraemic individuals were found relatively infrequently [24,25]. Even in plasma factor concentrates, produced from thousands of blood donations, genotype 2 DNA was only detected in a very small percentage of individuals [26]. In contrast, genotype 2 DNA was found at a much higher frequency in tissue from individuals older than approximately 40 years of age, leading to the assumption that genotype 2 has widely disappeared from circulation [27]. Genotype 3 virus was shown to be endemic in Ghana, West Africa [3] and may be present in a certain region of Brazil [28]. Outside these areas, only a few sporadic cases of viraemic infection have been reported in France [9] and one case was identified in the UK [15]. In an investigation performed in the USA, even the persisting genotype 3 DNA has only been found in a small percentage of tissue samples [29].

One of the main questions raised by the discovery of the new viral variants is whether the similarity between the genotypes results in restriction of infection to one genotype or whether multiple infections are possible.

Long-term persistence of PB19V DNA in human tissue is a well known feature [27-29]. After acute infection, residual viral DNA can remain in tissue for decades or even lifelong ('bioportfolio') [18]. Herein, we report on the co-persistence of different genotypes and several genomic variants belonging to the same genotype.

The mechanism by which persistence of several variants of genomes belonging to the same genotype is accomplished is currently unknown. Analogous to simultaneous persistence of two genotypes, reinfection might occur. However, since there was evidence for co-persistence of three or more variants, repeated reinfection has to be assumed. Another explanation could be that the genetic diversity has evolved during the acute infection. Especially during the initial phase of B19V infection, characterized by high levels of viral replication, there might be a greater chance of appearance of inaccurately replicated B19V DNA; although, the viral DNA is replicated by the host cell machinery. It has been recently speculated by others [30,; 31] that the fidelity and the proofreading activity of the enzymic complex containing the host DNA polymerase, recruited cellular replication factors and the viral NS1 protein may not be as efficient in single-stranded DNA viruses as expected so far. Furthermore, it has been demonstrated that PB19, similar to parvoviruses infecting animals, has a high rate of evolutionary changes that is more typical of RNA viruses [31, 31]. Since the NS1 gene evolves at a similar rate as the gene for VP2, it was suggested that immune selection is not the primary cause of the high substitution rate in PB19.

Several studies focused attention on the analysis of genetic variability in selected regions of the PB19 virus genome. Mainly three different genomic regions have been analysed and can be compared with the consensus sequence obtained from the global alignment: the promoter region, extending from within the left terminal region to the start of the NS reading region; the NS reading region, located in the left half of the genome; the VP reading region, located in the right half of the genome, and divided into a VP1 unique region and a VP1/VP2 common region. Investigations lead to some conclusions that can be as follows. (1) The promoter region may be more freely variable, however it should maintain intact its functional characteristics otherwise virus could loose its infectivity. (2) The NS region shows a limited degree of sequence variability and this may cause variability in the functional characteristics of NS protein, and therefore, of the biological properties of the virus. (3) The VP1 unique region shows a variable domain, corresponding to an exposed domain of the protein where many neutralising linear epitopes are present, and a conserved domain corresponding to the viral phospholipase domain, necessary for viral infectivity and for various pathological activity of the virus as well be discussed later on (4) The VP1/2 common region, composing the core of the capsid and also presenting many different conformational neutralizing epitopes, is less but more uniformly variable. Therefore, although of limited extent, variability in PB19 virus genome may account for different biological characteristics of the virus, but any hypothesis should be tested in adequate experimental systems.

In overall sequence, these three types differ from each other by 10%. The most striking variation is observed within the promoter area, in which the three virus types differ by >20%. Within the NS1 gene, sequence divergences between genotypes 2 and 3 and genotype 1 are 13% at the nucleotide level and 6% at the amino acid level. Within the open reading frame encoding the VP1/2 proteins, the majority of nucleotide substitutions are synonymous: at the nucleotide level, genotypes 2 and 3 differ from the prototype by 9 and 12%, respectively, but at the amino acid level they differ by only 1.1 and 1.4%. However, the degree of amino acid divergence within the VP1 unique region (uVP1) is higher: genotypes 2 and 3 differ from genotype 1 by 4.4 and 6.6%, respectively. Interestingly, amino acids 130 to 195 of the uVP1 gene containing the reported phospholipase 2 activity [8,32) are highly conserved, and variation is mostly clustered in the N termini. Since important neutralizing epitopes are located within this region, differences in antibody response/recognition might ensue. Although a high degree of antigenic cross-reactivity has been shown between genotypes 1 and 3, almost no data has been available on the corresponding immunological relationship between genotypes 1 and 2.

Post infection, the DNA of the PB19 prototype persists in solid tissues as an intact, continuous molecule devoid of any apparent persistence-specific mutations in the coding sequence (15) Furthermore, in our recent studies with over 500 samples of skin, tonsil, synovial, and liver tissues, the persistence of virus type 1 and 2 DNAs was shown to be frequent and lifelong [18], whereas persistent type 3 DNA was undetected in northern Europe. However, type 3 DNA has been encountered in blood endemically in Ghana and infrequently in France and Brazil [3, 9, 35, 33]. Our recent studies (18) suggest that in northern Europe both virus types 1 and 2 circulated widely until the 1960s, after which type 2 disappeared and has subsequently occurred only sporadically. The genome substitution rate of B19 type 1, similar to that of canine parvovirus 2 (CPV2) [34], has been shown to be very high, approaching those of many RNA viruses [35]. It is not known if the disappearance of virus type 2 was due to stochastic variation, to immunological differences, or to retardation of biological proficiency or fitness in comparison to that of the type 1 virus.

PB19 genotype 1 replicates restrictively in the erythrocyte precursors of bone marrow, yielding high viral loads in blood during the early stage of infection [36, 37,38,39]. In epidemiological studies with plasma pools from 100,000 Danish and 140,000 Finnish blood donors [40, 41] and studies of sera from 1,640 symptomatic patients, the DNA of PB19 virus type 1, but not of type 2 or 3, was detected [18]. By contrast, among blood donors in Ghana, the DNA of virus type 3 occurred more frequently but in low copy numbers [3]. Overall, high-titer viremias of virus types 2 and 3 have been encountered only sporadically [42, 43].

Interesting finding denotes that PB19 DNA of type 1 has been shown to persist in numerous tissue types, including synovial tissue, skin, liver, brain, muscle, and myocardial tissue, even though the mechanism of persistence and the cell type(s) involved are still unknown[44]. Correspondingly persistent DNA of virus type 2 was encountered initially mainly in skin, pointing to a possible difference in tissue tropism between the PB19 virus types [10]. However, studies of >500 tissues of various types revealed DNA of virus type 2, like that of the prototype, in all organs studied [18]. Furthermore, for both of the PB19 types, the persistence seemed to be life long. Moreover, in northern and central Europe, virus types 1 and 2 appear to have circulated equally until the 1960s, after which virus type 2 disappeared. By contrast, virus type 3 appears to have been absent from wide circulation in these areas during the past 70 years [18].

The molecular basis for these epidemiological differences is still unknown, as are, in general, the biological functions and full pathogenetic potentials of virus types 2 and 3. Nguyen et al. studied the infectivity of a virus type 2 isolate in UT7/Epo-S1 cells. No transcripts could be detected with RT-PCR. In contrast, Blümel et al. [42] observed equal mRNA expression of virus types 1 and 2 in KU812Ep6 cells. Until now, the biological activity of virus type 3 has not been studied. According to the the difference in the epidemiologies of the three PB19 types, there may be functional and immunological difference in the characteristics of the three virus types. The self-assembly of structural proteins into capsids has been shown for genotypes 1 [46, 47] and 3 [48]. Also the same feature has been found for genotype 2 with the capsid proteins VP2 and VP1/2 expressed in the baculovirus system. Electron micrographs revealed icosahedral particles with the same diameter (23 nm) as virus type 1 [49], which, furthermore, were able to hemagglutinate human erythrocytes. Under the same conditions, the type 1 to 3 viremic sera were shown, in addition to DNA, to contain virus particles with red cell surface-binding activity. The infectivity of these particles was demonstrated in two cell lines permissive for the parvovirus prototype. In both cell types, all three virus types induced transcription and maturation (splicing) of mRNAs and the synthesis of NS1 and VP proteins.

Another biological feature that has been associated with the host cell tropism of PB19 is the relative production rates of the structural and nonstructural proteins in the infected cells [50, 51]. This balance is determined by differential polyadenylation and splicing of the mRNA transcripts. Even though the internal polyadenylation sites, (pA)p1 and (pA)p2, of B19 pre-mRNAs are conserved between the three PB19 virus genotypes, it has been observed that there is differences within the recently defined downstream and upstream *cis*-acting elements and the adjacent B motif influencing the activity of this polyadenylation site [51]. Similarly, the sequence of the 11-kDa protein, which was recently shown to be critical for capsid protein production in the infected cells [52], differs between the three genotypes. These molecular divergences might contribute to the different epidemiologies [18], e.g., by influencing the tissue tropism and the persistence pattern. However, in that case this influence would need to be subtle, because it was found that both variant genotypes to produce capsid proteins and to infect the PB19 type 1-permissive cells. Furthermore, both variants have been associated with anemia or aplastic crisis, indicating erythroid tropism [9, 11, 44].

On the other hand, the specific absence of virus type 2 from the tissues of the young [18] could be argued to be due to an immunological difference between the PB19 types. In that scenario, a hypothetical mechanism for the maintenance of the genotype 1 would be equilibrium between viral replication and the efficiency of host immunity. However, phylogenetic analyses have shown the PB19 nucleotide substitutions to be mostly synonymous and the substitution rates of the VP1 and NS1 genes to be similar, suggesting that immune selection is not the primary driving force of PB19 virus evolution [36].

Virus types 1 to 3 differ, at the protein level, by only 1 to 2% within the capsid, although most variation resides within the unique portion of the minor capsid protein VP1, which contains important neutralizing epitopes. Studies of antigenic relationships between PB19 types 1 and 3 [48, 53] have shown a high degree of cross-reactivity, and in vitro neutralization tests using sera from subjects infected with PB19 type 1 have shown inhibition of in vitro infection of type 2 [42] . However, due to a general lack of methods to recognize type 2 IgG-positive sera and of type 2 recombinant antigens, until now it has not been possible to test whether sera from subjects infected with type 2 would react with antigens of type 1 or whether type 2 antigens are recognized equally well by type 1 sera. Previous data has shown 100% cross-reactivity in both VP2 and VP1/2 antibody EIAs between PB19 types 1 and 2 was seen. Such knowledge of cross-reactivity is relevant for both diagnosis and vaccine development, as well as for definition of the taxonomic status of these human erythrovirus variants [54]. DNA of PB19 type 1 has been shown to persist in numerous tissue types, including synovial tissue, skin, liver, brain, muscle, and myocardial tissue, even though the mechanism of persistence and the cell type(s) involved are still unknown. Correspondingly persistent DNA of virus type 2 was encountered initially mainly in skin, pointing to a possible difference in tissue tropism between the PB19 virus types [10].

Promoter Region

There is data concerning the sequence of promoter region from 11 different isolates obtained from serum of patients with acute or chronic manifestations of PB19 infections. When compared with the consensus sequence, ten variable positions were identified; five base changes were common to different isolates, four base changes were unique for single isolates, and a single base insertion was present in one isolate. Furthermore, no mutations could be identified in sequence motifs recognized by cellular transcription factors. Although no experimental data were presented in this study, the activity of the different promoter sequences was considered to be unaffected by single-base mutations and there was no relation between persistent infection and mutations [55].

Zakrzewska et al. (2001) analyzed the sequence of PB19 genome isolates obtained from synovial tissues of patients suffering from hemophilic arthritis and in 12 isolates obtained from serum in the course of acute infection as a control group. When compared with the consensus sequence, a total of 47 variable positions were identified; 22 were common to different isolates, and of these six showed the presence of three different possible bases, while the remaining 25 were unique for single isolates. Of the 47 variations, eight were in the terminal repeat region, 27 were in the unique region before the transcription start site and the remaining 12 were in the 5′UTR region of NS protein. Mutation clusters were particularly evident in the promoter region; however, they do not directly affect binding sites for transcription factors. The overall degree of genetic heterogeneity found in this study was relatively high, up to 4.0% with respect to the consensus sequence, but no differences were present in variability rates between the chronic arthropaties and acute infections groups. On the other hand, the distribution of the mutations showed that some mutations were typical of the arthropathies or acute infections groups, respectively, therefore, not excluding the hypothesis that persistence in synovial tissue might be correlated with a different transcriptional activity of the viral promoter.

NS Region

Ishii et al. (1999) submitted a group of 25 sequences in the nucleotide database of NCBI. Among this group there were three almost complete sequences of PB19 genome. The isolates were mainly obtained from serum in blood donors or different clinical situations; four isolates were obtained form synovial tissues in patients with rheumatoid arthritis. When compared with the consensus sequence, a total of 53 variable positions were found; of these, 28 were common to more than one isolate and 25 were unique for single isolates. The genetic heterogeneity in this group of isolates was quite diverse. However, these more variable isolates shared the same base polymorphisms common to all other isolates.

Hemauer et al. (1996) analyzed the sequence of a segment of PB19 genome in the NS C-terminal region in 20 different B19 isolates obtained from serum of patients with acute or chronic B19 virus infection. When compared with the consensus sequence, 30 variable positions were found in this region, ten common to more than one isolate and 20 unique for single isolates. The degree of genetic heterogeneity ranged from 0.22 to 1.78%, and was higher in isolates obtained from persistent infections with respect to isolates obtained from acute infections.

This variation in base sequence was reflected in variation at the amino acid level, where the degree of heterogeneity ranged from 0 to 4.0%, with higher values shown by isolates obtained from persistent infections with respect to isolates obtained from acute infections. These changes might correlate with different biological activities of NS protein: the authors in particular pointed out a substitution at amino acid 508 of NS protein, which in turn might influence the clinical course and immune reactivity in persistent infections.

4.4. VP Region

Sequence heterogeneity in the VP region was studied extensively by Erdman et al. (1996) in an effort to understand the geographical distribution and transmission patterns of PB19 virus and to identify genetic markers that may account for the diversity of clinical manifestations of infection.

Sequence was obtained for the VP region of multiple PB19 virus isolates obtained from viremic sera. Nineteen isolates were obtained in the USA between 1984 and 1994. Seven of these were from different communities and 12 were from a single community outbreak, four of these as convalescent-phase samples. Ten other isolates were obtained from different countries in Europe, Asia and South America. A consensus sequence was determined and heterogeneity was assessed encompassing the whole VP region.

When analyzed with respect to the consensus sequence obtained in this study, a total of 215 positions showed changes at the nucleotide level; 35 positions showed changes at the amino acid level and in two cases multiple amino acid changes were present. Nucleotide substitutions were evenly distributed, with the exception of a highly conserved region. An important correlation emerged from these data: conserved regions corresponded to the viral phospholipase domain in the VP1 unique region and to the core of capsid proteins, while variable regions corresponded to regions exposed on the capsid surface, important for the presence of linear or conformational epitopes highly immunogenic and relevant for neutralization of infectivity.

The sequences analyzed in this study differed from the consensus sequence from 2 to 99 nucleotides and from 0 to 13 amino acids. Pair-wise comparison permitted the construction of a dendrogram by neighbor-joining method. The highest degree of homology was shown by isolates obtained from single outbreaks, in particular from related isolates. High similarity was shown between isolates from the community outbreak in Ohio when compared with other isolates from USA; two isolates from China formed a cluster, more related to a single isolate from Japan. Paired isolates showed conserved sequences, only a single isolate showed a silent nucleotide change. Nucleotide differences might suggest geographic or temporal clustering, however, evidence showed that multiple lineages might co circulate in the same epidemiological settings. In this study once again, no sequence pattern or genotypic grouping could be associated with particular clinical presentations.

In the study of Hemauer et al. (1996) sequence variability was analyzed also in the VP1 N terminal region, the VP2 N terminal region and the VP1/2 common C-terminal region. Sequence data were analyzed and compared with a consensus sequence. In the VP1 N-terminal region, variability was equally distributed over the whole sequence at both nucleotide and amino acid levels, in contrast to other data, and ranged from 0.3 to 4% at the nucleotide level and from 0 to 1.7% at the amino acid level. Two isolates from persistently infected patients showed a higher degree of variation (6.8 and 8.2%, respectively).

No distinction was evident between the immunogenic and phospholipase domains in this VP region. In the VP2 N-terminal region, data showed the presence of a conserved section, where only three nucleotide substitutions were present in 185 positions, followed by a highly variable section, where 133 substitutions were present in 230 positions.

The overall variability ranged from 1.2 to 1.7%, higher in two isolates obtained from persistent infections (2.4 and 2.9%, respectively). At the amino acid level, fewer positions were involved in changes resulting in a 0.9% variability (1.8% in isolates form persistent infections). In the VP1/2 C-terminal region, variability was distributed over the whole region, with values ranging from 0 to 2.4% at the DNA level and from 0 to 3.6% at the protein level. More than 50% of isolates did not show amino acid changes.

Conclusions from the authors were that isolates from persistently infected subjects showed a higher degree of variability both at the DNA and protein levels, this being possibly explained by the continuous replication, accumulation of changes and selection for amino acid alterations, which in some cases were identically conserved. This might be related to selective pressure for functional proteins or particular immunogenic epitopes; otherwise, this may simply reflect epidemiological clustering of viral isolates [60].

The variability of genomic sequence in smaller segments of the VP region has been investigated by other groups. Gallinella et al. (1995) reported a low degree of variability in a 1000 nt segment from the N-terminus of VP proteins in different isolates obtained from sera of patients with different clinical manifestations of B19 infection.

In this group of isolates, 15 nucleotide changes were identified with respect to a consensus sequence, seven common to more than one isolate and eight specific for single isolates. The number of variations per isolate ranged from 2 to 6, the average genetic distance among isolates being 0.63%, while at the amino acid level there were from zero to three substitutions per isolate.

In the study by Takahashi et al. (1999), two different genomic regions were analyzed: segment from the N-terminus of VP1 (VP1 unique region, and a segment from the N-terminus of VP2 (VP1/VP2 junction region. Six isolates were obtained from sera in the course of an outbreak of PB19 virus infection in a pediatric surgery ward and were strictly correlated as regards the transmission of infection. The variability detected in the VP unique region ranged from the 0.29 to the 0.87% at the DNA level and from the 0.43 to the 2.17% at the protein level. Inspection against a consensus derived from lull-length sequences showed the presence of three polymorphic sites located in the N terminus of VP1 protein. The variability detected in the VP1/2 common region was lower and evenly distributed. Of particular interest in this study were the demonstration of the identity of viral sequences between two isolates suggesting direct transmission of the virus and, on the other hand, the detection of a mixed viral population in one infected patient, suggesting either coinfection by multiple strains or mutation due to replication in the host. However, no correlation emerged between viral sequence and clinical manifestations.

Sequence data for the VP1 N-terminal region have been submitted in the nucleotide database of NCBI also by Tsay and Hsu 1999 and Sol-Church et al. 2001.

Dorsch et al. (2001) analyzed a segment in the VP1 unique region and N terminal region of VP2 in 16 different isolates obtained from sera of nine acutely infected pregnant women with different clinical presentation, five arthritis and two chronically infected subjects. Variability was found in 24 nucleotide and 21 amino acid positions.

The amino acids in the regions 18-136 and 189-247 were more affected by changes, while the conserved domain of viral phospholipase was not affected by changes. The degree of variability at the amino acid level ranged from 0 to 2%, and was higher (3.2%) in one isolate from a chronic infection. The sequences of isolates obtained from fetuses were identical to those of maternal isolates, except in one case, where a mutation was identified and supposed to have occurred in the fetus. In this work, selected variants were tested for affinity towards human monoclonal antibodies directed against a localised epitope in the VP1 unique region. Amino acid sequence variations did not affect the binding affinities of human monoclonal antibodies, suggesting that sequence variation did not affect the structural conformation of this epitope, located at amino acid 30-42 and probably presented at the surface of the virion in an accessible and conserved conformation.

Human parvovirus B19-VP1u is known to have a phospholipase A2 (PLA2) motif [67-69] and its enzyme activity has been associated with various inflammatory processes that may contribute to the activation of macrophage. Activity of sPLA2 has been implicated in a variety of physiological and pathological responses, including cell proliferation, chemokinesis [70, 71], ECM remodeling [72-75], vascular inflammation, [76-77] and cerebral ischemia [78]. Recently, parvovirus B19-VP1u has been linked with sPLA2-like activity that is recognized as group XIII enzyme [67-69], a novel type of sPLA2 that may contribute to various pathological processes [79-81]. Additionally, a PB19-VP1u mutant, PB19-VP1uD175A, is mutated in the phospholipase domain and loses its enzymatic activity and viral infectivity [82-84]. Parvovirus B19-VP1u and its sPLA2 enzymatic activity are critical for eliciting macrophage responses associated with a variety of inflammatory processes. Our experimental results demonstrate the effects of sPLA2 activity in PB19-VP1u proteins by increasing migration, phagocytosis, and inflammatory responses such as significant increases of MMP9 activity, interleukin 6 (IL-6) and interleukin 1 β IL-1β mRNA expression in macrophages. This study strongly suggests the crucial role of sPLA2 activity of PB19-VP1u in macrophage activation and may provide clues in understanding the role of PB19-VP1u in the host response to PB19 infection and PB19-related diseases.

VARIANT GENOTYPES OF PARVOVIRUS B19

Nguyen et al. (1998, 1999) reported the identification of a variant isolate, detected in 1995 in the serum of a child presenting with aplastic anaemia in G6PD deficiency, that showed marked sequence divergence from the consensus B19 sequence in a 346 nt segment in the VP1 unique region. Base sequence in this segment was more than 11% divergent from the sequence of other PB19 isolates, and it was clearly unrelated to the sequence of other animal erythroviruses. The complete genome of the variant isolate, defined as V9, was then cloned and its sequence determined (Nguyen, 1999), confirming that sequence divergence was not limited at the VP1 unique region but extended to the whole genome. The detection of this variant isolate opened the issue whether other PB19-related variant isolates could be present in the human population.

In fact, a high level of sequence divergence may affect the molecular identification of erythrovirus isolates in the routine diagnosis of erythrovirus infection as usually conducted by means of PCR, DNA hybridisation or serological assays. The findings of variant erythrovirus isolates might extend the etiological role of erythroviruses in human diseases and help to elucidate the degree of circulation of these viruses in the population.

Furthermore, the role of acquired immunity directed against PB19 virus in the protection against related viruses might in turn affect the epidemiology of these variant erythroviruses.

On this basis, several groups developed molecular assays apt to detect and distinguish both classical, B19-like sequences, and variant, V9-like sequences. Heegaard et al. (2002) failed to detect variant isolates in the screening of 100 anti-B19 IgM positive serum samples and of 50 plasma pools, each composed by 2000 single donations, obtained from danish blood donors.

In the study conducted by Nguyen et al. (2002) analysis involved testing, by using a consensus PCR assay able to differentiate B19 and variant isolates, of: (a) serum samples submitted to NIH between 1991 and 1998, previously screened and found positive by dot-blot hybridization using a PB19-specific probe; (b) serum samples submitted to NIH from 1998 to 2001; (c) 62 plasma pools, each composed by 2000 single donations, obtained from danish blood donors.

All samples found positive by the assay were PB19 virus positive, with the exception of one single sample, submitted for testing in 1991 from Italy and collected from an HIV infected individual presenting with chronic anemia. The entire coding region of the isolate, defined as A6, was cloned and sequence was determined from two independently derived clones (clones were 99.7% identical).

The isolate was as divergent both from PB19 isolates (87.8% similarity) and V9 isolate (92.0%similarity) as B19 isolates and V9 were between them (86.6% similarity). The divergence between A6 and B19 was less pronounced at the protein level (similarity of 94.2% in NS region, 93.8% in VP1 unique region, 96.7% in VP proteins region). As this was the only variant isolate found in a large screening program, questions were raised regarding the real epidemiological relevance of variant erythroviruses; however, it was noted that more divergent isolates, while being detected by dot-blot hybridisation, might be undetected by PCR assays not specifically designed.

A different approach was followed by Hokynar et al. (2002). PB19 virus had previously been detected in long-term persistence in bone marrow, synovia and skin. While isolates obtained from synovia were quite similar to those obtained from peripheral blood or bone marrow, new variant viruses emerged when samples of skin were analysed. Paired skin and serum samples were available from 34 subjects, 19 of them were seropositive to PB19 and from 14 of these viral DNA was detected by using a PCR assay specific for the VP1 unique region. Of these, five could also be amplified by using other standard primer sets, corresponding to a classical PB19 virus sequence, while nine were only amplified with this primer set suggesting the presence of either fragmented genomes or variant isolates.

Two viral genomes were cloned and sequenced, showing the presence of variant isolates. Primer sets specifically designed also showed that all non-PB19 DNA from skin could be assigned to this new group of variant isolates. The sequences obtained form two isolates (LaLi and HaAM) were only 0.3% divergent from each other, but widely divergent from classical PB19 sequences. The divergence was high in the promoter region, showing a 26.5% divergence from PB19 sequence, especially deletions; the promoter was, however, functional in reporter assays. Sequence divergence was lower in the coding regions: 12.9, 4.6, 12.9% at the nucleotide level, 6.0, 4.4, 1.1% at the amino acid level in the NS, VP1 unique, VP2 regions, respectively.

Overall, no PB19 seronegative persons had viral DNA, while 14 out 19 B19 seropositive persons had either B19 (n-5) or variant sequences (n-9) isolated form their skins. Variant isolates were more prevalent in skin samples, but they could not be found in samples from bone marrow or synovia. This finding, for the first time, identifies a clear correlation between a viral genotype and tissue tropism in the human erythrovirus group.

The most comprehensive study has so far been conducted by Servant et al. (2002) to identify variant erythroviruses, to evaluate the possible circulation and the clinical presentation of variant viruses, and to specify the taxonomic groupings of these viruses. For this purpose, a consensus PCR was designed and employed for the detection and discrimination, by means of restriction site polymorphism analysis, between viral genomes.

More than 1000 serum or plasma samples were analysed, divided in several groups on the basis of epidemiological and clinical settings. (a) HIV patients, France, 1992-1997; (b) foetal hydrops, France, 1995-1997; (c) PB19 related symptoms or exposed pregnant women, USA; (d) PB19 virus antigen positive, France, 1972-1999; (e) prospective study, France, 1999-2001.

A total of 396 out of 1084 samples were found positive to the consensus PCR assay. Of these, 385 had a PB19-type restriction pattern, while 11 had a different, V9-type restriction pattern. The relative frequency differed among different groups: V9 pattern was not identified in samples from USA, while 9 of the 11 isolates came from samples of the prospective group from France. No statistically significant differences were found between the prevalence in the different sample groups from France. Thus, V9 related viruses circulated among the French population and could be responsible for symptomatic infections clinically indistinguishable from classical PB19 infection: aplastic crisis, chronic anaemia, rash, pancytopenia, and foetal hydrops.

To assess the genetic diversity among V9 related viruses, base sequence was determined for a full length clone of a variant viral isolate (D91.1). The overall genomic arrangement and organization were similar for PB19 and V9 isolates. Sequence alignment could, therefore, be performed to analyse the genetic diversity between these viruses. Taken together, the data from this study identified three different clusters among erythrovirus isolates.

This was confirmed by further investigating from six variant isolates. Sequence alignment and phylogenetic analysis clearly corroborated the presence of three genetic clusters within the human erythrovirus group. It has been proposed that the human erythroviruses should be divided, on the basis of genetic distances and evolutionary relationships, in three genotypes: genotype I corresponding to PB19-related isolates, genotype II to LaLi-related isolates and genotype III to V9-related isolates.

The identification of variant isolates within the human erythrovirus group led to the division of human erythroviruses in three distinct genotypes and identification of three phylogenetic lineages.

The high ratio of synonymous to non synonymous nucleotide substitutions might be considered indicative of ancient separation between lineages, and this view is corroborated by the finding of multiple isolates belonging to different lineages.

However, the origin and the epidemiological relevance of these viruses are still obscure, especially considering the limited analysis conducted so far for determining the prevalence of variant isolates. Furthermore, the molecular assays used so far may not be adequate for the sensitive detection of variant viruses of variable, unknown sequences and, although the pathological and clinical presentations of infection seem not different among PB19 genotypes, this may hamper the detection of variants viruses in unexpected clinical situations and their correlation with particular clinical manifestations. At the antigenic level, the viral variants show extended cross-reactivity and baculovirus-expressed antigens may be used in assays that will detect antibodies directed toward all genotypes. However, the VP1 unique region contains, beside a conserved phospholipase domain, a variable immunogenic region containing most VP1 linear neutralising epitopes. Thus, subtle differences in the immune response and in the degree of cross-neutralisation should be considered as important factors in the epidemiology and clinical course of erythrovirus infections.

Conclusion

Genetic analysis of parvovirus B19 has been carried out mainly to establish a framework to track molecular epidemiology of the virus and to correlate sequence variability with different pathological and clinical manifestations of the virus. Good information regarding PB19 virus sequence variability is now available. However, most studies showed that the genetic variability of PB19 virus is low, that molecular epidemiology is possible only on a limited geographical and temporal setting, and that no clear correlations are present between genome sequence and distinctive pathological and clinical manifestations. More recently, several viral isolates have been identified. There has been an identification of three divergent genetic clusters, about 10% divergent from each other and still quite distinct from other parvoviruses, that can be thought of as different genotypes within the human erythrovirus group and that show clearly resolved phylogenetic relationship. These variant isolates pose interesting questions regarding the real extent of genetic variability in the human erythroviruses, and the relevance of these viruses in terms of epidemiology and their possible implication in the pathogenesis of erythrovirus-related diseases.

References

[1] Heegaard E D & Brown K E. Human parvovirus B19. *Clin. Microbiol. Rev.* 2002, 15: 485–505.

[2] Kerr JR. Pathogenesis of parvovirus B19 infection: host gene variability, and possible means and effects of virus persistence. *J. Vet. Med. B Infect. Dis. Vet. Public Health.* 2005, 52: 335–339.

[3] Candotti D, Etiz N, Parsyan A & Allain J P. Identification and characterization of persistent human erythrovirus infection in blood donor samples. *J. Virol.* 2004, 78: 12169–12178

[4] Heegaard E D, Petersen B L, Heilmann C J & Hornsleth A. Prevalence of parvovirus B19 and parvovirus V9 DNA and antibodies in paired bone marrow and serum samples from healthy individuals. *J. Clin. Microbiol.* 2002, 40: 933–936

[5] Lefrere J J, Servant-Delmas A, Candotti D, Mariotti M, Thomas I., Brossard Y, Lefrere F, Girot R, Allain J P & Laperche S. Persistent B19 infection in immunocompetent individuals: implications for transfusion safety. *Blood.* 2005, 106: 2890–2895

[6] Siegl G, Cassinotti P. Parvoviruses. In: Mahy BWJ, Collier L, editors. Topley & Wilson's Microbiology and Microbial Infection, vol. 1, ninth ed. London: Arnold, 1998:261-79.

[7] Chipman PR, Agbandje-McKenna M, Kajigaya S, Brown KE, Young NS, Baker TS, Rossmann MG. Cryo-electron microscopy studies of empty capsids of human parvovirus B19 complexed with its cellular receptor. *Proc. Natl. Acad. Sci. U.S.A.* 1996;93:7502-6.

[8] Dorsch S, Liebisch G, Kaufmann B, von Landenberg P, Hoffmann JH, Drobnik W, Modrow S. The VP1 unique region of parvovirus B19 and its constituent phospholipase A2-like activity. *J. Virol.* 2002; 76:2014-17

[9] Servant A, Laperche S, Lallemand F, Marinho V, De Saint Maur G, Meritet J F & Garbarg-Chenon A. Genetic diversity within human erythroviruses: identification of three genotypes. *J. Virol.* 2002, 76, 9124–9134

[10] Hokynar K M, Soderlund-Venermo M, Pesonen A, Ranki O, Kiviluoto E K, Partio and Hedman K. A new parvovirus genotype persistent in human skin. *Virology.* 2002, 302:221 228

[11] Nguyen Q T, Wong S, Heegaard E D, and Brown K E. Identification and characterization of a second novel human erythrovirus variant, A6. *Virol.* 2002, 301:374-380

[12] Parsyan, A., C. Szmaragd, J. P. Allain, and D. Candotti. Identification and genetic diversity of two human parvovirus B19 genotype 3 subtypes. *J. Gen. Virol.* 2007, 88:428-431.

[13] Cohen B J, Gandhi J, and Clewley J P. Genetic variants of parvovirus B19 identified in the United Kingdom: implications for diagnostic testing. *J. Clin. Virol.* 2006, 36:152-155

[14] Eis-Hubinger A M, Reber U, Abdul-Nour T, Glatzel U, Lauschke H, and Putz U. Evidence for persistence of parvovirus B19 DNA in livers of adults. *J. Med. Virol.* 2001, 65:395-401.

[15] Hokynar K, Brunstein J, Soderlund-Venermo M, Kiviluoto O, Partio E K, Konttinen Y, and Hedman K. Integrity and full coding sequence of B19 virus DNA persisting in human synovial tissue. *J. Gen. Virol.* 2000, 81:1017-1025.

[16] Kerr J R, Cartron J P, Curran M D, Moore J E, Elliott J R, Mollan R A. A study of the role of parvovirus B19 in rheumatoid arthritis. *Br. J. Rheumatol.* 1995, 34:809-813

[17] Kuhl U, Pauschinger M, Noutsias M, Seeberg B, Bock T, Lassner D, Poller W, Kandolf R, Schultheiss H P. High prevalence of viral genomes and multiple viral infections in the myocardium of adults with "idiopathic" left ventricular dysfunction. *Circulation.* 2005, 111:887-893.

[18] Norja P, Hokynar K, Aaltonen L M, Chen R, Ranki A, Partio E K, Kiviluoto et al. Bioportfolio: lifelong persistence of variant and prototypic erythrovirus DNA genomes in human tissue. *Proc. Natl. Acad. Sci. U.S.A.* 2006, 103:7450-7453.[

[19] Soderlund M, Essen R V, Haapasaari J, Kiistala U, Kiviluoto O, Hedman K. Persistence of parvovirus B19 DNA in synovial membranes of young patients with and without chronic arthropathy. *Lancet.* 1997, 349:1063-1065

[20] Manning A, Willey S J, Bell J E, Simmonds P. Comparison of tissue distribution, persistence, and molecular epidemiology of parvovirus B19 and novel human parvoviruses PARV4 and human bocavirus. *J. Infect. Dis.* 2007, 195:1345-1352

[21] Bergallo M, Costa C, Sidoti F, Novelli M, Ponti R, Castagnoli C, Merlino C, Bernengo MG, Cavallo R. Variants of parvovirus B19: bioinformatical evaluation of nested PCR assays. *Intervirol.* 2008; 51(2):75-80.

[22] Blumel J, Eis-Hubinger A M, Stuhler A, Bonsch C, Gessner M & Lower J Characterization of parvovirus B19 genotype 2 in KU812Ep6 cells. *J. Virol.* 2005, 79, 14197–14206

[23] Liefeldt L, Plentz A, Klempa B, Kershaw O, Endres A S, Raab U, Neumayer H H, Meisel H, Modrow S. Recurrent high level parvovirus B19/genotype 2 viremia in a renal transplant recipient analyzed by real-time PCR for simultaneous detection of genotypes 1 to 3. *J. Med. Virol.* 2005, 75:161-169

[24] Schneider B, Hone A, Tolba R H, Fischer H P, Blumel J, Eis-Hubinger A M. Simultaneous persistence of multiple genome variants of human parvovirus B19. *J. Gen. Virol.* 2008, 89:164-176.

[25] Sanabani S, Neto W K, Pereira J, and Sabino E C. Sequence variability of human erythroviruses present in bone marrow of Brazilian patients with various parvovirus B19-related hematological symptoms. *J. Clin. Microbiol.* 2006, 44:604-606.

[26] Wong S, Young N S & Brown K E. Prevalence of parvovirus B19 in liver tissue: no association with fulminant hepatitis or hepatitis-associated aplastic anemia. *J. Infect. Dis.* 2003, 187, 1581–1586

[27] Cassinotti P, Burtonboy G, Fopp M. & Siegl G. Evidence for persistence of human parvovirus B19 DNA in bone marrow. *J. Med. Virol.* 1997, 53, 229–232

[28] Eis-Hübinger A M, Reber U, Abdul-Nour T, Glatzel U, Lauschke H. & Pütz, U. Evidence for persistence of parvovirus B19 DNA in livers of adults. *J. Med. Virol.* 2001, 65, 395–401

[29] Soderlund M, von Essen R, Haapasaari J, Kiistala U, Kiviluoto O, Hedman K. Persistence of parvovirus B19 DNA in synovial membranes of young patients with and without chronic arthropathy. *Lancet.* 1997, 349:1063-1065.

[30] López-Bueno A, Mateu M G & Almendral J M. High mutant frequency in populations of a DNA virus allows evasion from antibody therapy in an immunodeficient host. *J. Virol.* 2003,77, 2701–2708

[31] Shackelton L A & Holmes E C. Phylogenetic evidence for the rapid evolution of human B19 erythrovirus. *J. Virol.* 2006, 80, 3666–3669.

[32] Zadori Z, Szelei J, Lacoste M C, Gariepy Y Li S, Raymond P, Allaire M, Nabi I R, Tijssen P. A viral phospholipase A2 is required for parvovirus infectivity. *Dev. Cell.* 2001, 1:291-302

[33] Nguyen Q T,Sifer C, Schneider V, Allaume X, Servant A, Bernaudin F, Auguste V, Garbarg-Chenon A. Novel human erythrovirus associated with transient aplastic anemia. *J. Clin. Microbiol.* 1999, 37:2483-2487

[34] Shackelton L A, Parrish C R, Truyen U, Holmes E C. High rate of viral evolution associated with the emergence of carnivore parvovirus. *Proc. Natl. Acad. Sci. U.S.A.* 2005, 102:379-384

[35] Ekman A, Hokynar K, Kakkola L, Kantola K, Hedman L, Bondén H, Gessner M, Aberham C, Norja P, Miettinen S, Hedman K, Söderlund-Venermo M. Biological and Immunological Relations among Human Parvovirus B19 Genotypes 1 to 3 *J. of Virol.* 2007, 81, 13, 6927-6935.

[36] Anderson M J, Jones S E, Minson A C. Diagnosis of human parvovirus infection by dot-blot hybridization using cloned viral DNA. *J. Med. Virol.* 1985, 15:163-172

[37] Enders M, Schalasta G, Baisch C, Weidner A, Pukkila L, Kaikkonen L, Lankinen H, Hedman L, Söderlund-Venermo M, Hedman K. Human parvovirus B19 infection during pregnancy—value of modern molecular and serological diagnostics. *J. Clin. Virol.* 2006, 35:400-406.

[38] Ozawa K, Kurtzman G, Young N. Replication of the B19 parvovirus in human bone marrow cell cultures. *Science.* 1986, 233:883-886

[39] Srivastava A, and Lu L. Replication of B19 parvovirus in highly enriched hematopoietic progenitor cells from normal human bone marrow. *J. Virol.* 1988, 62:3059-3063.

[40] Heegaard E D, Panum Jensen I, and Christensen J. Novel PCR assay for differential detection and screening of erythrovirus B19 and erythrovirus V9. *J. Med. Virol.* 2001, 65:362-367.

[41] Hokynar K, Norja P, Laitinen H, Palomaki P, Garbarg-Chenon A, Ranki A, Hedman K, Söderlund-Venermo M. Detection and differentiation of human parvovirus variants by commercial quantitative real-time PCR tests. *J. Clin. Microbiol.* 2004, 42:2013-2019.

[42] Blümel J, Eis-Hübinger A M, Stühler A, Bönsch C, Gessner M, Löwer J. Characterization of parvovirus B19 genotype 2 in KU812Ep6 cells. *J. Virol.* 2005, 79:14197-14206

[43] Liefeldt L, Plentz A, Klempa B, Kershaw O, Endres A S, Raab U, Neumayer H H, Meisel H, Modrow S. Recurrent high level parvovirus B19/genotype 2 viremia in a renal transplant recipient analyzed by real-time PCR for simultaneous detection of genotypes 1 to 3. *J. Med. Virol.* 2005, 75:161-169

[44] Nguyen Q T, Sifer C, Schneider V, Bernaudin F, Auguste V, Garbarg-Chenon A. Detection of an erythrovirus sequence distinct from B19 in a child with acute anaemia. *Lancet.* 1998, 352: 9139/ 1524-1524

[45] Söderlund-Venermo M, Hokynar K, Nieminen J, Rautakorpi H, Hedman K. Persistence of human parvovirus B19 in human tissues. *Pathol. Biol.* 2002, 50:307-316

[46] Brown C S, Van Lent J W, Vlak J M, Spaan W J. Assembly of empty capsids by using baculovirus recombinants expressing human parvovirus B19 structural proteins. *J. Virol.* 1991, 65:2702-2706.

[47] Kajigaya S, Fujii H, Field A, Anderson S, Rosenfeld S, Anderson L J, Shimada T, and Young NS. Self-assembled B19 parvovirus capsids, produced in a baculovirus system, are antigenically and immunogenically similar to native virions. *Proc. Natl. Acad. Sci. U.S.A.* 1991, 88:4646-4650

[48] Heegaard ED, Qvortrup K, and Christensen J. Baculovirus expression of erythrovirus V9 capsids and screening by ELISA: serologic cross-reactivity with erythrovirus B19. *J. Med. Virol.* 2002, 66:246-252.

[49] Cossart Y E, Field A M, Cant B, and Widdows D. Parvovirus-like particles in human sera. *Lancet.* 1975, i: 72-73.

[50] Liu J M, Green S W, Shimada T, and Young N S. A block in full-length transcript maturation in cells nonpermissive for B19 parvovirus. *J. Virol.* 1992, 66:4686-4692

[51] Yoto Y, Qiu J, and Pintel D J. Identification and characterization of two internal cleavage and polyadenylation sites of parvovirus B19 RNA. *J. Virol.* 2006. 80:1604-1609

[52] Zhi N, Mills I P, Lu J, Wong S, Filippone C, and Brown K E. Molecular and functional analyses of a human parvovirus B19 infectious clone demonstrates essential roles for NS1, VP1, and the 11-kilodalton protein in virus replication and infectivity. *J. Virol.* 2006, 80:5941-5950

[53] Parsyan A, Kerr S, Owusu-Ofori S, Elliott G & Allain, J P Reactivity of genotype-specific recombinant proteins of human erythrovirus B19 with plasmas from areas where genotype 1 or 3 is endemic. *J. Clin. Microbiol.* 2006, 44, 1367–1375

[54] Tattersall, P. The evolution of parvovirus taxonomy, 5-14. In J. R. Kerr, S. F. Cotmore, M. E. Bloom, R. M. Linden and C. R. Parrish (ed.), 2006, Parvoviruses, vol. 1. Hodder Arnold, London, United Kingdom

[55] Gareus R, Gigler A, Hemauer A, Leruez-Ville M, Morinet F, Wolf H, Modrow S. Characterization of cis-acting and NS1 protein-responsive elements in the p6 promoter of parvovirus B19. *J. Virol.* 1998; 72:609_/16.

[56] Zakrzewska K, Azzi A, De Biasi E, Radossi P, De Santis R, Davoli PG, Tagariello G. Persistence of parvovirus B19 DNA in synovium of patients with haemophilic arthritis. *J. Med. Virol.* 2001; 65:402-7.

[57] Ishii KK, Munakata Y, Funato T, Fu Y, Koseki N, Sugamura K, Sasaki T. Sequence of human parvovirus B19 isolates from patients with rheumatoid arthritis. 1999. NCBI Nucleotide Database, AB030673-AB030694, unpublished.

[58] Hemauer A, von Poblotzki A, Gigler A, Cassinotti P, Siegl G, Wolf H, Modrow S. Sequence variability among different, parvovirus B19 isolates. *J. Gen. Virol.* 1996; 77:1781-5.

[59] Erdman DD, Durigon EL, Wang QY, Anderson LJ. Genetic diversity of human parvovirus B19: sequence analysis of the VP1/VP2 gene from multiple isolates. *J. Gen. Virol.* 1996;77:2767-74.

[60] Gallinella G, Venturoli S., Manaresi E, Musiani M. & Zerbini M. B19 virus genome diversity: epidemiological and clinical correlations. *J. Clin. Virol.* 2003, 28, 1–13.

[61] Gallinella G, Venturoli S, Gentilomi G, Musiani M, Zerbini M. Extent of sequence variability in a genomic region coding for capsid proteins of B19 parvovirus. *Arch. Virol.* 1995; 140:1119-25.

[62] Takahashi N, Takada N, Hashimoto T, Okamoto T. Genetic heterogeneity of the immunogenic viral capsid protein region of human parvovirus B19 isolates obtained from an outbreak in a pediatric ward. *FEBS Lett.* 1999; 450:289_/93.

[63] Tsay GJ, Hsu TC. Sequence variability of human Parvovirus B19 found in patients with Parvovirus B19 infection and systemic lupus erythematosus. 1999. NCBI Nucleotide Database, AF217236_/ AF217244, unpublished.

[64] Sol-Church K, Simpson-Small T, Tung J, Mason RW. Incidence of Human Parvovirus B19 in a Pediatric Population. 2001. NCBI Nucleotide Database, AF380251_/AF380256, unpublished.

[65] Dorsch S, Kaufmann B, Schaible U, Prohaska E, Wolf H, Modrow S. The VP1-unique region of parvovirus B19: amino acid variability and antigenic stability. *J. Gen. Virol.* 2001;82:191_/9.

[66] Zádori Z, Szelei J, Lacoste MC, Li Y, Gariépy S, Raymond P, Allaire M, Nabi IR, Tijssen P. A viral phospholipase A2 is required for parvovirus infectivity. Dev. Cell. 2001;1:291–302.

[67] Canaan S, Zádori Z, Ghomashchi F, Bollinger J, Sadilek M, Moreau ME, Tijssen P, Gelb MH. Interfacial enzymology of parvovirus phospholipases A2. J. Biol. Chem. 2004;279:14502–14508.

[68] Lu J, Zhi N, Wong S, Brown KF. Activation of Synoviocytes by the Secreted Phospholipase A2 Motif in the VP1-Unique Region of Parvovirus B19 Minor Capsid Protein. J Infect Dis. 2006;193:582–590.

[69] Filippone C, Zhi N, Wong S, Lu J, Kajigaya S, Gallinella G, Kakkola L, Söderlund-Venermo M, Young NS, Brown KE. VP1u phospholipase activity is critical for infectivity of full-length parvovirus B19 genomic clones. *Virology*. 2008; 374:444–452.

[70] Kerr J R, Cartron J P, Curran M D, Moore J E, Elliott J R, and Mollan R A. A study of the role of parvovirus B19 in rheumatoid arthritis. *Br. J. Rheumatol.* 1995, 34:809-813.

[71] Kuhl, U, Pauschinger M, Noutsias M, Seeberg B, Bock T, Lassner D, Poller W, Kandolf R, and Schultheiss H P. High prevalence of viral genomes and multiple viral infections in the myocardium of adults with "idiopathic" left ventricular dysfunction. *Circulation*. 2005, 111:887-893.

[72] Arita H, Hanasaki K, Nakano T, Oka S, Teraoka H, Matsumoto K. Novel proliferative effect of phospholipase A2 in Swiss 3T3 cells via specific binding site. *J. Biol. Chem.* 1991;266:19139–19141.

[73] Choi YA, Lim HK, Kim JR, Lee CH, Kim YJ, Kang SS, Baek SH. Group IB secretory phospholipase A2 promotes matrix metalloproteinase-2-mediated cell migration via the phosphatidylinositol 3-kinase and Akt pathway. J. Biol. Chem. 2004;279:36579–36585.

[74] Lee CH, Lee J, Choi YA, Kang SS, Baek SH. cAMP elevating agents suppress secretory phospholipase A2-induced matrix metalloproteinase-2 activation. *Biochem. Biophys. Res. Commun.* 2006;340:1278–1283.

[75] Hiromi II, Naoya H, Issei T, Mayuko O, Takashi S, Satoshi A. Group IVA phospholipase A2-associated production of MMP-9 in macrophages and formation of atherosclerotic lesions. *Biol. Pharm. Bull.* 2008;31:363–368.

[76] Oestvang J, Johansen B. PhospholipaseA2: a key regulator of inflammatory signalling and a connector to fibrosis development in atherosclerosis. *Biochim. Biophys. Acta.* 2006;1761:1309–1316.

[77] Hurt-Camejo E, Camejo G, Peilot H, Oörni K, Kovanen P. Phospholipase A(2) in vascular disease. *Circ. Res.* 2001;89:298–304.

[78] Vikman P, Ansar S, Henriksson M, Stenman E, Edvinsson L. Cerebral ischemia induces transcription of inflammatory and extracellular-matrix-related genes in rat cerebral arteries. *Exp. Brain Res.* 2007;183:499–510.

[79] Oestvang J, Johansen B. PhospholipaseA2: a key regulator of inflammatory signalling and a connector to fibrosis development in atherosclerosis. *Biochim. Biophys. Acta.* 2006;1761:1309–1316.

[80] Hurt-Camejo E, Camejo G, Peilot H, Oörni K, Kovanen P. Phospholipase A(2) in vascular disease. *Circ. Res.* 2001;89:298–304.

[81] Vikman P, Ansar S, Henriksson M, Stenman E, Edvinsson L. Cerebral ischemia induces transcription of inflammatory and extracellular-matrix-related genes in rat cerebral arteries. *Exp. Brain Res.* 2007; 183:499–510.

[82] Brennan FM, Feldmann M. Cytokines in autoimmunity. *Curr. Opin. Immunol.* 1996;8:872–877.

[83] MacNaul KL, Hutchinson NI, Parsons JN, Bayne EK, Tocci MJ. Analysis of IL-1 and TNF-alpha gene expression in human rheumatoid synoviocytes and normal monocytes by in situ hybridization. *J. Immunol.* 1990;145:4154–4166.

[84] Attiga FA, Fernandez PM, Weeraratna AT, Manyak MJ, Patierno SR. Inhibitors of prostaglandin synthesis inhibit human prostate tumor cell invasiveness and reduce the release of matrix metalloproteinases. *Cancer Res.* 2000;60:4629–4637.

Chapter 8

LABORATORY DIAGNOSIS OF PB19

ABSTRACT

PB19 can cause acute, generally self-limiting diseases in immunocompetent children and adults. Data suggests that the infection also has been associated with a wide variety of clinical manifestations and some clinical features of B19 infection, such as anemia, arthropathy and rash. The infection leads to a viremia that can be present, at high titer, for about one week, followed by the onset of a specific immune response controls the infection. B19 infection in pregnancy can be associated with non-immunologic foetal hydrops or foetal death. In immunocompromised hosts, B19 can persist over several months and sometimes years. Persistent or recurrent B19 infections can be associated with chronic clinical manifestations or with transient clinical syndromes, generally related to the recrudescence of viral replication. An important issue is that screening of blood donors and blood products may be needed in some times to avoid complications to pregnant and to immunocompromised patients. So a laboratory diagnosis of B19 infection is required. A diagnostic protocol must consider both the type of pathology and the type of patient. In immunocompetent individuals serological and virological testing is complementary, while in immunocompromised patients viral detection is the diagnosis of choice.

Keywords: PB19, viral antigen detection, molecular method, serology.

Introduction

Increased recognition of PB19 as a significant human pathogen that causes fetal loss and severe disease in immunocompromised patients has resulted in intensive efforts to understand the pathogenesis of PB19-related disease, to improve diagnostic strategy that is necessary to detect PB19 infection and blood-product contamination and, finally, to explain the nature of the cellular immune response that is elicited by the virus [1].

Understanding of the clinical presentations and pathogenesis of infection will facilitate diagnosis. The first phase of PB19 infection is characterized by viremia that develops approximately 6 days after intranasal inoculation of PB19 into susceptible individuals who lack serum antibodies to the virus. The viremia lasts for 2-3 days through which period of viral detection is at best detection level from samples of nasopharyngeal secretions, serum, peripheral leucocytes, amniotic fluid and tissue lysate. Multiple samples are usually preferred due to the short period of viremia related to the clinical manifestations of the disease. The appearance of PB19 specific IgM antibodies in the serum in the second week after inoculation corresponds with clearance of the viremia. In the third week after inoculation, specific IgG antibodies appear in the serum [2].

Diagnosis of PB19 Infection

A diagnostic protocol must consider both the type of pathology and the type of patient. In immunocompetent individuals serological and virological testing is complementary, while in immunocompromised patients viral detection is the only feasible diagnostic approach. Viral detection methods are generally based nowadays on the direct detection of PB19 genome in clinical specimens. According to these considerations; we can plan the diagnosis of PB19 in the following steps [3].

1) Diagnostic Cytopathology

2) Direct detection of the virus by:

- Electron microscopy (EM).
- Culture of the virus.

- Detection of PB19 antigens in the serum.
- Immunohistological detection of PB19 antigen in formalin-fixed Paraffin-embedded tissue biopsy.
- DNA hybridization or Polymerase Chain Reaction (PCR).

3) Detection of Specific Antibodies to PB19 Structural Proteins VP1, VP2 & Non-Structural Protein NS1

1) Diagnostic Cytopathology (Figure 5)

The cytopathic effect of infection of erythroid progenitor cells with PB19, both in vivo and in vitro, is manifested as giant pronormoblasts (named as lantern cells), first recognized in 1948 in the BM of patients with transient aplastic crisis. Giant pronormoblasts are early erythroid cells with a diameter of 25 to 32 μm, large eosinophilic nuclear inclusion bodies, and cytoplasmic vacuolization, and occasionally, "dog-ear" projections may be observed [4]. Electron microscope (EM) of cells reveals cytopathic ultrastructural changes that include pseudopod formation, marginated chromatin, and virus particles in the nucleus [2].

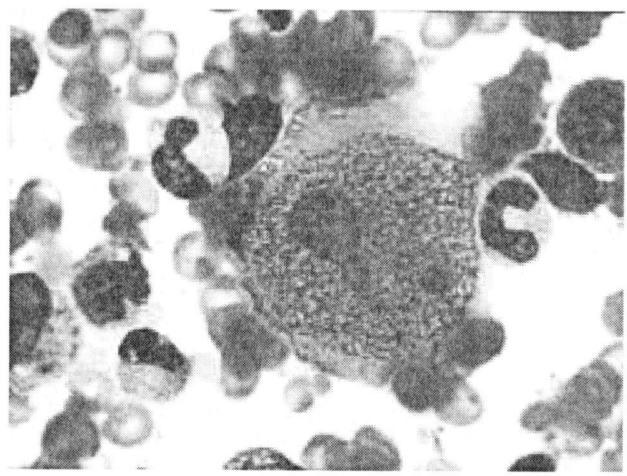

Figure (6): Degenerated giant proerythroblast (May- Grünwald-Giemsa staining) **[4].**

It is noteworthy that the presence of giant pronormoblasts in either BM or peripheral blood is only suggestive of PB19 infection. Since these cells are often absent in patients with human immunodeficiency virus (HIV) infection or other chronic infections; the cytopathology is of limited diagnostic value [2].

2) Direct Detection of the Virus by

a) Electron Microscope (EM)

The EM can under optimal conditions provide the quickest diagnosis as the time taken may be as short as half an hour, but this is true only when relatively large concentrations (10^6/ml) of the virus are relatively accessible in the patient. Antibodies, alone or combined with a marker such as colloidal gold, can enhance the sensitivity of EM [2].

The PB19 virion appears by EM as a simple structure composed of only two proteins and a linear, single-stranded DNA molecule. The non-enveloped viral particles are 22 to 24 nm in diameter and show icosahedral symmetry, and often both empty and full capsids are visible by negative staining and EM [5]. Mature infectious viral particles have a molecular weight of 5.6 x 10^6 Dalton [6].

EM is used also to study the cytopathic effect of PB19 on erythroid progenitors. These ultrastructural changes include margination of chromatin, pseudeopod formation, vaculation and capsid particles in lacunae within nuclear chromatin [7]. The detection of these changes and the presence of viral particles in fetal tissue during the viremic phase are suggestive of PB19 viral infection [8].

b) Culture of PB19

There is no animal model for PB19, and virus can only be grown in culture with difficulty. PB19 can be cultured in erythroid progenitor cells from a variety of sources, including human BM, fetal liver, cord blood, and peripheral blood.

It was found that PB19 (B19) tropism is almost restricted to erythroid progenitor cells and mediated by receptor and coreceptor interactions: i.e., P antigen [9, 10], $\alpha_5\beta_1$ integrin complex [11], and Ku80 antigen [12]. In vitro replication of PB19 is limited to few models that allow production of PB19 infectious particles and study of the virus life cycle.

The first used cell line is that one involved the use of bone marrow erythroid cells [13-15]. Furthermore the use of cell line, derived from an erythroleukemia clone [16, 17] called UT7, has been generally employed [18-23]; however, this allow limited replication at a level that cannot allow mass production of PB19. Other erythroleukemia cell lines have been used [24-26], without demonstrating their clear usefulness in B19 production [27]. Thus, the major source of virus is currently obtained from sera of viremic patients.

The data described by Wong et al. represent PB19 replication from CD36$^+$ erythroid progenitors [28] obtained from mobilized blood and claims to be the first description of an ex vivo method to produce large numbers of erythroid progenitor cells that are highly permissive to PB19 infection and replication.

Another cell type from CD36$^+$ erythroid progenitor cells were generated by Freyssinier et al. in 1999 from CD34$^+$ peripheral adult or cord blood stem cells [29]. The cells consisted of 96% late erythroid burst-forming units and erythroid colony-forming cells: i.e., the main in vivo targets of PB19. After infection of these CD36$^+$ erythroid progenitor cells with PB19 infectious particles, it has showed the interaction of NS1 protein with the tumor necrosis factor alpha pathway leading to apoptosis of the infected erythroid progenitor cells [16]. Infection of CD36$^+$ cells with PB19 also allowed expression of VP1 and VP2 capsid proteins in about 50% of the cells 24 hours post infection [16,30]. Better expression of PB19 protein was enhanced when CD36$^+$ erythroid cells under hypoxia were cell line was exposed to low oxygen tension [31]. Hypoxia increases PB19 replication, leading to the production of higher titers of PB19 infectious particles. This phenomenon resulted, at least in part, from an interaction of the major cellular factor implicated in oxygen tension response, hypoxia-inducible factor 1, with the P6 promoter. All of these studies that showed the usefulness of CD36$^+$ erythroid progenitor cells to produce B19 infectious particles as well as to be a relevant model to explore B19 infectious cycle should have been taken into account.

Other cell line was used also for PB19 propagation such as: two megakaryoblastoid cell lines [32, 33] and two human erythroid leukemia cell lines [34]. These lines have been used to study mechanisms of replication. Moreover, in vitro studies of PB19 in explanted human bone marrow (BM) cultures have confirmed the erythroid specificity of this virus. In spite all of this, the yield of virus from all these cultures is poor, and they can't be used as a source of antigen for diagnostic tests [2].

An important fact that in all culture systems erythropoietin is required to maintain viral replication, probably by supporting the rapid division of erythroid progenitors. All systems are culture for short time only and are not suitable for long-term culture [2].

c) Detection of PB19 Antigens in the Serum

PB19 viral antigens include the two capsid proteins VP1 & VP2 which are encoded by genes on the right side of PB19 DNA genome. Each capsid consists of an icosahedral structure with a total of 60 capsomers: VP2 is the major structural protein, accounting for 96% of total capsid protein. The minor capsid protein, VP1, is identical to VP2 with the addition of 227 amino acids (termed the VP1 unique region) at the amino terminus[2]. VP2 and VP1 can be expressed in bacterial, mammalian, and insect cells. In mammalian and insect cells, expression of VP2 can self-assemble in the absence of viral DNA to produce virus-like particles (VLP) that are physically, antigenically and immunogenically similar to native virions [35]. Although a divergence of up to 3% on the amino acid level has been noted in different PB19 strains, there is no evidence of more than one antigenic strain [2].

There is a simple and rapid dot immunoperoxidase assay for the direct detection of PB19 capsid antigens in human sera. The assay is performed with serum specimens dotted onto nylon membranes. VP1 and VP2 PB19 antigens are detected with a pool of monoclonal antibodies directed against the two proteins, and the complex is visualized by immunoperoxidase staining. The assay could be performed in about 4 hours, and positive results are revealed at the end of the reaction as dark blue spots on the nylon membrane at the site of positive specimens [36].

This method is applied to large number of serum samples and the results obtained by the dot immunoperoxidase assay were compared with the results obtained for the presence of PB19 DNA by dot blot hybridization and nested PCR. With optimized working conditions, the dot immunoperoxidase assay was able to detect the presence of PB19 with sensitivity comparable or slightly higher than that achieved by dot blot hybridization but less than that achieved by nested PCR. Since the level of sensitivity of the dot immunoperoxidase assay proved to be appropriate for the detection of acute PB19 infection, and since the cost, turnaround time, and versatility of the assay are important issues, the dot immunoperoxidase assay described may be particularly suitable for large-scale screening of samples and a good alternative to DNA detection methods in the routine laboratory evaluation of PB19 infection [36].

Another methods used for detection of PB19 viral antigens are the Enzyme Linked Immunosorbant Assay (ELISA) & Radioimmunoassay (RIA) techniques. In ELISA, the assay is based upon class specific anti-PB19 antibodies bound to a solid phase component. The antigens present in patient's samples bind to these specific antibodies and forms an immune complex. The enzyme conjugate attaches to this complex, then after adding the substrate, a color is produced by the bound enzyme. The intensity of the color is directly proportional to the amount of antigen present in the samples. This technique is highly specific & sensitive.

The principle of RIA resembles that of ELISA, but a radioactive isotope is employed as the marker for labeling anti-viral antibodies. Undoubtedly, this method is very sensitive & specific particularly in patients with impaired immunity.

D) Immunohistochemical Detection of PB19 Antigen in Formalin-Fixed

Paraffin-embedded tissue biopsy:

Immunohistochemical techniques (IHC) are used as a quick, easy and sensitive method for clinical diagnosis. It is one of the widely used techniques for viral detection, tumor diagnosis, detecting prognostic markers and predicting therapeutic response. IHC techniques permit the demonstration of specific cell and tissue antigens in tissue sections, cell smears and cytospins using enzymes or fluorochromes as the label. This method is used to detect PB19 VP1 &VP2 in a variety of tissues, but most commonly in fetal tissues, liver tissue & BM biopsy sections [37].Detection systems are immune-enzymatic, immunogold or immunoflouorescent. Recently, many developments in the amplification of signals by the addition of primary antibodies, biotinylated secondary antibodies, streptavidin-biotin peroxidase complex, alkaline phosphatase anti-alkaline phosphatase complexes and peroxidase-conjugated rabbit anti-human antibody enhanced polymer staining system.

In the avidin - biotin complex (ABC) system, tissue sections are incubated with a primary antibody to the tissue antigen of interest. A biotinylated secondary antibody is then added, and followed by the addition of avidin. The final step then, is to add a biotin enzyme indicator complex. The tissue antigen is demonstrated by cytochemical staining of the enzyme indicator in the complex. The enzyme indicator most commonly used is horseradish peroxidase. Other enzymes such as alkaline phosphatase or glucose oxidase have also been used [7].

E) Detection of Viral DNA

The need to monitor PB19 viremias in immunocompromised patients, to test the efficacy of immunoglobulin therapy and to evaluate viral clearance has prompted the development of a quantitative PCR & DNA hybridization assay to detect PB19 DNA in different samples [37, 38].

Viral gene amplification by polymerase chain reaction PCR which is a powerful technique for detecting minute amounts of PB19 DNA, but at the risk of specimen contamination; it is more prone to false positive results than other diagnostic tests due to cross contamination of the specimen or reaction mixture with extraneous viral sequences. To overcome this problem; PCR with double primer pairs (nested PCR) is done.

Nested PCR implies that two pairs of PCR primers are used for a single locus. The first pair amplifies the locus and the second pair of primers (nested primers) binds within the first PCR product and produces a second PCR product that will be shorter than the first one. The logic behind this strategy is that if the wrong locus were amplified by mistake, the probability is very low that it would also amplified a second time by a second pair of primers.

The most sensitive tests are nested PCR systems, but these assays provide semiquantitative results at best. A parvovirus B19 DNA assay was developed based on the real time TaqMan PCR. This method was calibrated on the basis of serial plasmid dilutions and tested with an international parvovirus B19 standard. The assay was capable of quantifying parvovirus B19 DNA from one to about 5 x 10(7) genome equivalents per reaction (corresponding to 100 to 5 x 10(9) genome equivalents per ml serum) [39].

For patients in early infection viremia occurs with extremely high titers (up to 10^{11} to 10^{13} genome equivalents/ml) (40, 41). At this point the patients tend to be asymptomatic, and B19-infected blood donors easily go unrecognized. The high-titer viremia lasts for only about 5 to 10 days, until expression of specific antibodies. When the virus is thought to be cleared from blood, its DNA at low levels can be detected by PCR in a declining proportion for months or, less frequently, years after infection [42, 43]. This level needs to be detected by PCR in certain situations concerning blood and blood proucts transfusion [44]. Plasma pools also contain antibodies. However, in solvent-detergent plasma safety studies [45], the recipients were shown to seroconvert due to pools containing B19 DNA in high titers (10^7 IU/ml). These patients also became B19 DNA positive, verifying virus transmission. In contrast, transfusion of B19 DNA in concentrations of $<10^4$ IU/ml is not considered to result in detectable seroconversion, with some exceptions (2). Consequently, according to the revised European Pharmacopoeia [46], plasma pools used for human anti-D immunoglobulin production, must not (after 1 January 2004) contain $>10^4$ IU/ml of B19 DNA. In order to identify such high-titer units, a quantitative DNA detection method is required. On the other hand, new variants of the PB19 virus have been discovered [47-50], leading to segregation of the species into three distinct genotypes [51] diverging from each other in sequence by >10%. In consequence, B19 PCR methods might fail to detect such variants due to sequence divergence between primers and viral target DNA.

Real-time PCR assay was developed with lower detection limit for quantitative detection of B19 DNA in clinical serum samples. The assay was carried out using a LightCycler instrument and product formation was monitored continuously with the fluorescent double-stranded DNA binding dye SYBR Green I. With an optimized PCR protocol, this system was able to quantitate the target DNA down to 3 x 10(1) genome copies/reaction and to detect as few as 3 x 10(0) genome copies/reaction [52].

PCR is a very sensitive method for detecting PB19 DNA in urine, amniotic fluid, pleural fluid, synovial fluid and serum. It can reveal PB19 DNA in patients who have not yet developed specific IgM and IgG antibodies and in some sera where specific IgM and/or IgG are present. PCR detection of PB19 in amniotic fluid is promising for identification of fetuses at high risk [39].

On the other hand, direct hybridization as a southern blot or dot blot format, generally employs a full length viral DNA probe labeled with P^{32}, digoxigenin, or biotin to bind to DNA in clinical specimens. Results of the hybridization assay are readily quantifiable, with a detection limit of ~10^5 genome copies/ml, and the hybridization assay will detect all known variants of PB19, including V9 [2]. The nucleic acid hybridization assay can detect the medium to high levels of viremia (above 10^4 genome copies of PB19 DNA) which occur in the acute phase of PB19 infection, and the PCR assay can detect even very low viral titers (between 1 and 100 PB19 genome copies). The search for PB19 by DNA hybridization techniques, however, is limited to certain laboratories,

Although direct hybridization is sensitive enough to detect PB19 levels in acute transient aplastic crisis and pure red cell aplasia due to PB19 infection; in immunosuppressed patients lower levels of viremia will be missed. The advent of PCR has greatly increased the sensitivity of DNA detection in serum and tissue samples, although it possesses a great propensity for contamination. DNA may be detectable for extended periods of time in serum, synovial membranes, and BM, even in healthy individuals. Therefore, the presence of low levels of PB19 DNA alone cannot be used to diagnose acute PB19 infection [2,53] .

The use of appropriate samples is needed for accurate diagnosis. In diagnosis of hydropes fetalis the fetal cord blood and amniotic fluid sample for detection of PB19 DNA is the most specific method. This highlights that it is suggested that an accurate laboratory diagnosis of PB19 infection cannot always rely on maternal samples but rather on virological analysis of fetal samples. Both the fetal cord blood and amniotic fluid sample are suitable for diagnosis, but the detection of

PB19 DNA in amniotic fluid samples by in situ hybridization proved to be the most reliable diagnostic system [54].

3) Detection of Specific Antibodies to PB19 Structural Proteins VP1, VP2 and Non-Structural Protein NS1

In most people the virus is highly immunogenic, and at the time of presentation almost all patients have detectable IgM antibodies, which persist for 2 to 3 months after acute infection. Antibody production is correlated with the disappearance of the virus from the bloodstream, and IgG production appears to confer lasting immunity [55].

Laboratory diagnosis of PB19 infection can be built on the basis of serologic testing (IgM & IgG) antibodies in immunocompetent patients; while DNA assays may be required in immunocompromised patients who are unable to amount an antibody response [55]. Also presence of giant pronormoblasts in the BM or peripheral blood may be suggestive of PB19 infection [2].

Although PB19 DNA-based assays are crucial for the diagnosis of PB19 infection presenting as transient aplastic crisis (before the antibody response) and in chronic infections in immunosuppressed individuals (who fail to mount an immune response), diagnosis of PB19 infection in immunocompetent individuals presenting with erythema infectiosum or PB19-induced arthropathy is by detection of PB19-specific antibody [2].

In a trial to improve the diagnostic value of parvovirus IgM in pregnant patients, we evaluated a cutoff value that diagnosed all parvovirus- positive RSA with PCR. The value was 78.5±30.12 IU/mL compared with 30.02 ±17.64 IU/mL in cases with PCR negative for parvovirus. There was a statistically significant difference ($P > .001$). This can be used as a simple approach to predict cases with parvovirus viremia. Another approach was plotting the receiver operator characteristic to find the value with improved specificity of parvovirus IgM. The value was 48 IU/mL, with 83.3% sensitivity, 66.7% specificity, and 89.2% accuracy [56].

All patients develop IgG antibodies against the capsid proteins VP1 & VP2, the majority of virus neutralizing antibodies that offer life-long protection against reinfection are directed against the VP1 unique region. IgM antibodies are mainly directed against VP2- specific epitopes. These antibodies can be present for only a rather short period of 2-10 weeks after acute infection. IgG antibodies against NS1 are preferentially found in patients who are unable to eliminate the virus and develop persisting viremia or virus persistence in distinct organs, eg. synovial fluid, liver & BM [57].

The PB19 IgG prevalence seems to vary in different countries eg. About 25% of all adults in Germany are IgG positive for PB19, whereas an IgG positivity of about 50% was found among pregnant women in USA.

Serological diagnosis of PB19 infection is generally achieved by detection of IgM & IgG antibodies to PB19 structural proteins VP1&VP2 and NS1. Detection is most often performed by the following methods:

- ELISA & RIA are mainly used for IgM antibody detection.
- ELISA & indirect Immunoflourescent Assay Test (IFA) are mainly used for IgG antibody detection. [3].
- ELISA and western blotting employing recombinant NS1 protein as antigen are mainly used for detection of PB19 NS1-specific antibodies [58]. Due to the inability of PB19 to efficiently replicate in culture systems, viral capsids were initially purified from serum with high virus titer and used for antibody tests. PB19 antigen can be expressed in bacteria, in cell lines, or as peptides, but currently most antigen is produced in insect cell lines with recombinant baculovirus. These recombinant antigens are noninfectious, and serologic results correlate well with those using native virus. Commercial assays are dependent upon expression of capsid proteins in the baculovirus expression system, which has proven to be very efficient in producing large quantities of empty VP1/VP2 and VP2 capsids [2].

Söderlund et al., (1995), estimated that IgG antibodies to linear epitopes of VP2 disappear at approximately 6 months postinfection, leaving only those antibodies that recognize undenatured VP2. This loss of epitope recognition can be problematic when using Escherichia coli-based enzyme immunoassays (EIAs) for the detection of PB19-specific IgG antibodies because they employ only denatured antigen. Studies by Franssila et al., (1996) later supported the loss of epitope recognition finding as well. More recently it was observed that approximately 16% of serum samples that were confirmed to be positive for PB19 IgG did not react with denatured VP1 or denatured VP2. This lack of PB19 IgG reactivity with denatured epitopes presents a limitation to serological assays incorporating denatured PB19 antigens.

Based on these findings, both the source and nature of the viral antigen(s) used in a serological assay are important variables to consider in evaluating its performance. Capture EIAs employing native or recombinant antigens are the best choices for measuring these antibodies. Systems employing either an E. coli or baculovirus expression vector have been described and used to express PB19 capsid protein. E. coli-based expression vectors produce linear proteins. In contrast, baculovirus-based expression vectors produce conformational proteins. In fact, baculovirus-expressed VP2 (BVP2) has been shown to self-assemble into empty capsids whose appearance, as evidenced by electron microscopy, is very similar to that of the native PB19 virion capsids [61].

Conclusion

In conclusion, parvovirus B19 infection is a new, emerging infectious disease that has also been implicated in the pathogenesis of several autoimmune or inflammatory conditions and hyropes fetalis. The laboratory diagnosis of parvovirus PB19 is problematic due to false positive IgM and the presence of PCR-detectable PB19 DNA in healthy persons. The development of better techniques for diagnosis is imperative for the availability of effective therapy with intravenous IgG for severe disease and of intrauterine blood transfusions for prevention of hydrops in the fetus. Viral detection methods are generally based, nowadays, on the direct detection of B19 genome in clinical specimens. PB19 DNA is mainly detected by hybridizations assays and by the most sensitive PCR assays. Serological diagnosis of PB19 infection is generally achieved by detection of IgM and IgG antibodies to the PB19 structural proteins VP1 and VP2. IgM detection is most often performed by capture assays, both in EIA and RIA formats, IgG are mainly detected by indirect EIA and immunofluorescence tests.

References

[1] Corcoran A and Doyles S Advances in the biology, diagnosis & host pathogen interaction of parvovirus B19. *J. Med. Microbiol.* 2004, 53:459-475.

[2] Heegaard ED and Brown KE. Human Parvovirus B19. *Clin. Microb. Rev.* 2002, 15:485-505.

[3] Zerbini M, Gallinella G, Cricca M, et al. Diagnostic procedures in B19 infection. *Pathol. Biol.* 2002, 50:332-38.

[4] Koduri PR. Novel cytomorphology of the giant proerythroblasts of parvovirus B19 infection. *Am. J. Hematol.* 1998 , 58:95-99.

[5] Berns KI. Parvoviridae: the viruses and their replication, In BN Fields, DM Knipe, PM Howley, et al. Fields virology. 1996, Lippincott-Raven, Philadelphia, Pa.p: 2173-97.

[6] Kerr JR. The Parvoviridae; an emerging virus family. *Infect. Dis. Rev.* 2000, 2: 99-109.

[7] Morey AL, Ferguson D J P and Leslie KO Intracellular localization of parvovirus B19 nucleic acid at the ultra-structural level by in situ hybridization with digoxigenin-labelled probes. *Histochem.* 1993, 25: 421.

[8] Gallinella G, Anderson SM, Young NS, et al. Human parvovirus B19 can infect cynomolgus monkey marrow cells in tissue culture. *J. Virol.* 1995, 69:3897-3899.

[9] Brown, K. E., S. M. Anderson, and N. S. Young. 1993. Erythrocyte P antigen: cellular receptor for B19 parvovirus. *Science.* 262114-117.

[10] Brown, K. E., J. R. Hibbs, G. Gallinella, S. M. Anderson, E. D. Lehman, P. McCarthy, and N. S. Young.. Resistance to parvovirus B19 infection due to lack of virus receptor (erythrocyte P antigen). *NEJM.* 1994, 3:30/1192-1196

[11] Munakata Y, Saito-Ito T, Kumura-Ishii K, Huang J, Kodera T, Ishii T, Hirabayashi Y, Koyanagi Y, and Sasaki T. Ku80 autoantigen as a cellular coreceptor for human parvovirus B19 infection. *Blood.* 2005, 10:6/3449-3456.

[12] Weigel-Kelley K A, Yoder M C, and Srivastava A. $\alpha_5\beta_1$ integrin as a cellular co-receptor for human parvovirus B19: requirement of functional activation of β_1 integrin for viral entry. *Blood.* 2003, 10:2/3927-3933.

[13] Ozawa, K., G. Kurtzman, and N. Young. 1987. Productive infection by B19 parvovirus of human erythroid bone marrow cells in vitro. Blood 70384-391.

[14] Ozawa K, Kurtzman G, and Young N. Replication of the B19 parvovirus in human bone marrow cell cultures. *Science.* 1986, 23:3/883-886.

[15] Srivastava A, and Lu L. Replication of B19 parvovirus in highly enriched hematopoietic progenitor cells from normal human bone marrow. *J. Virol.* 1988, 623059-3063

[16] Sol N, Junter J L, Vassias I, Freyssinier J M, Thomas A, Prigent A F, Rudkin B B, Fichelson S, and Morinet F. Possible interactions between the NS-1 protein and tumor necrosis factor alpha pathways in erythroid cell apoptosis induced by human parvovirus B19. *J. Virol.* 1999, 73:8762-8770

[17] Shimomura S, Komatsu N, Frickhofen N, Anderson S, Kajigaya S, and Young N S. First continuous propagation of B19 parvovirus in a cell line. *Blood.* 1992, 79:18-24.

[18] Ekman, A., K. Hokynar, L. Kakkola, K. Kantola, L. Hedman, H. Bondén, M. Gessner, C. Aberham, P. Norja, S. Miettinen, K. Hedman, and M. Söderlund-Venermo. Biological and immunological relations among human parvovirus B19 genotypes 1 to 3. *J. Virol.* 2007, 81:6927-6935

[19] Gallinella G, Manaresi E, Zuffi E, Venturoli S, Bonsi L, Bagnara G P, Musiani M, Zerbini M. Different patterns restriction to B19 parvovirus replication in human blast cell lines. *J. Virol.* 2000, 27:8/361-367.

[20] Leruez, M., Pallier C, Vassias I, Elouet J F, Romeo P, and Morinet F. Differential transcription, without replication, of non-structural and structural genes of human parvovirus B19 in the UT7/EPO cell as demonstrated by in situ hybridization. *J. Gen. Virol.* 1994,75:1475-1478.

[21] Wong S, and Brown K. E. Development of an improved method of detection of infectious parvovirus B19. *J. Clin. Virol.* 2006, 35:407-413.

[22] Zhi, N., Mills I. P, Lu J, Wong S, Filippone C, and Brown K E. Molecular and functional analyses of a human parvovirus B19 infectious clone demonstrates essential roles for NS1, VP1, and the 11-kilodalton protein in virus replication and infectivity. *J. Virol.* 2006, 80:5941-5950.

[23] Zhi, N., Zadori Z., Brown K. E., and Tijssen P. Construction and sequencing of an infectious clone of the human parvovirus B19. *Virology.* 2004, 31:8/142-152

[24] Blümel, J., A. M. Eis-Hübinger, A. Stühler, C. Bönsch, M. Gessner, and J. Löwer. Characterization of parvovirus B19 genotype 2 in KU812Ep6 cells. *J. Virol.* 2005, 79:14197-14206.

[25] Miyagawa, E., Yoshida T., Takahashi H., Yamaguchi K., Nagano T., Kiriyama Y., Okochi K., and Sato H.. Infection of the erythroid cell line, KU812Ep6, with human parvovirus B19 and its application to titration of B19 infectivity. *J. Virol. Methods.* 1999, 8:345-54.

[26] Takahashi, T., Ozawa K., Takahashi K., Okuno Y., Muto Y., Takaku F., and Asano S. DNA replication of parvovirus B 19 in a human erythroid leukemia cell line (JK-1) in vitro. *Arch. Virol.* 1993, 13 :1, 201-208.

[27] Caillet-Fauquet P, Draps M L, Giambattista M Di, Launoit Y de, and Laub R. Hypoxia enables B19 erythrovirus to yield abundant infectious progeny in a pluripotent erythroid cell line. *J. Virol. Methods*. 2004, 12:1,145-153.

[28] Shade R O, Blundell M C, Cotmore S F, Tattershall P, and Astell C R. Nucleotide sequence and genome organization of human parvovirus B19 isolated from a child during aplastic crisis. *J. Virol.* 1986, 58:921-936

[29] Freyssinier, J. M., C. Lecoq-Lafon, S. Amsellem, F. Picard, R. Ducrocq, P. Mayeux, C. Lacombe, and S. Fichelson. Purification, amplification and characterization of a population of human erythroid progenitors. *Br. J. Haematol.* 1999, 106912-922

[30] Pillet, S., Z. Annan, S. Fichelson, and F. Morinet. Identification of a nonconventional motif necessary for the nuclear import of the human parvovirus B19 major capsid protein (VP2). *Virology*. 2003, 30:625-32.

[31] Pillet, S., N. Le Guyader, T. Hofer, F. NguyenKhac, M. Koken, J. T. Aubin, S. Fichelson, M. Gassmann, and F. Morinet. Hypoxia enhances human B19 erythrovirus gene expression in primary erythroid cells. *Virology*. 2004, 327:1,1-7

[32] Shimomura S, Komatsu N, Frickhofen N, et al. First continuous propagation of B19 parvovirus in a cell line. *Blood.* 1992, 79:18-24.

[33] Munshi NC, Zhou S, Woody MJ, et al. Successful replication of parvovirus B19 in the human megakaryocytic leukemia cell line MB-02. *J. Virol.* 1993, 67:562-66.

[34] Pillet S, Fichelson S, and Morinet F. Human B19 Erythrovirus in vitro replication: What's new? *J. Virol.* 2008; 82(17): 8951–8953

[35] Kajigaya S, Fujii H, Field A, et al. Self-assembled B19 parvovirus capsids, produced in a baculovirus system, are antigenically and immunogenically similar to native virions. *Proc. Natl. Acad. Sci. U.S.A.* 1991,88:4646-50.

[36] Gentilomi G, Musiani M, Zerbini M, et al. Dot immunoperoxidase assay for detection of parvovirus B19 antigens in serum samples. *J. Clin. Microbiol.* 1997,35: 1575–78.

[37] Pinho JRR, Alves VAF, Vieira AF et al. Detection of human parvovirus B19 in patients with hepatitis. *Baz. J. Med. Biol. Res.* 2001, 34:1131-38.

[38] Naber SP. Molecular Pathology -- Diagnosis of Infectious Disease. *NEJM.* 1994, 331:1212-15.

[39] Knoll A, Louwen F, Kochanowski B, et al. Parvoviruse B19 infection in pregnancy. Quantitative viral DNA analysis using kinetic detection method. *J. Med. Virol.* 2002, 67:259-66.

[40] Anderson, M. J., S. E. Jones, and A. C. Minson. Diagnosis of human parvovirus infection by dot blot hybridization using cloned viral DNA. *J. Med. Virol.* 1985,15:163-172

[41] Leruez, M., C. Pallier, I. Vassias, J. F. Elouet, P. Romeo, and F. Morinet. Differential transcription, without replication, of non-structural and genes of human parvovirus B19 in the UT7/EPO cell demonstrated by in situ hybridization. *J. Gen. Virol.* 1994, 75:1475-1478.

[42] Ekman, A., K. Hokynar, L. Kakkola, K. Kantola, L. Hedman, H. Bondén, M. Gessner, C. Aberham, P. Norja, S. Miettinen, K. Hedman, and M. Söderlund-Venermo. 2007. Biological and immunological relations among human parvovirus B19 genotypes 1 to 3. *J. Virol.* 81:6927-6935

[43] Srivastava, A., and L. Lu. Replication of B19 parvovirus in highly enriched hematopoietic progenitor cells from normal human bone marrow. *J. Virol.* 1988, 62:3059-3063

[44] Hokynar, K., M. Söderlund-Venermo, M. Pesonen, A. Ranki, O. Kiviluoto, E. K. Partio, and K. Hedman. 2002. A new parvovirus genotype persistent in human skin. *Virology.* 302:224-228.

[45] Shimomura, S., N. Komatsu, N. Frickhofen, S. Anderson, S. Kajigaya, and N. S. Young. First continuous propagation of B19 parvovirus in a cell line. *Blood.* 1992, 79:18-24

[46] Sugawara, H., R. Motokawa, H. Abe, M. Yamaguchi, Y. Yamada-Ohnishi, J. Hirayama, H. Sakata, S. Sato, N. Kamo, K. Ikebuchi, and H. Ikeda. Inactivation of parvovirus B19 in coagulation factor concentrates by UVC radiation: assessment by an in vitro infectivity assay using CFU-E derived from peripheral blood $CD34^+$ cells. *Transfusion.* 2001, 41:456-461.

[47] Shade, R. O., M. C. Blundell, S. F. Cotmore, P. Tattershall, and C. R. Astell. Nucleotide sequence and genome organization of human parvovirus B19 isolated from a child during aplastic crisis. *J. Virol.* 1986, 58:921-936.

[48] Nguyen, Q. T., C. Sifer, V. Schneider, F. Bernaudin, V. Auguste, and A. Garbarg-Chenon. Detection of an erythrovirus sequence distinct from B19 in a child with acute anemia. *Lancet.* 1998, 352:1524.

[49] Nguyen, Q. T., C. Sifer, V. Schneider, X. Allaume, A. Servant, F. Bernaudin, V. Auguste, and A. Garbarg-Chenon. Novel human erythrovirus associated with transient aplastic anemia. *J. Clin. Microbiol.* 1999, 37:2483-2487.

[50] Nguyen, Q. T., S. Wong, E. D. Heegaard, and K. E. Brown. Identification and characterization of a second novel human erythrovirus variant, A6. *Virology*. 2002, 301:374-380

[51] Servant, A., S. Laperche, F. Lallemand, V. Marinho, G. De Saint Maur, J. F. Meritet, and A. Garbarg-Chenon. Genetic diversity within human erythroviruses: identification of three genotypes. *J. Virol.* 2002, 76:9124-9134

[52] Manaresi E, Gallinella G, Zuffi E, Bonvicini F, Zerbini M, Musiani M. Diagnosis and quantitative evaluation of parvovirus B19 infections by real-time PCR in the in the clinical laboratory. *J. Med. Virol.* 2002 67; 2, 275 - 281

[53] Siegl G and Cassinotti P. Presence and significance of parvovirus B19 in blood and blood products. *Biologicals*. 1998 26:89-94.

[54] Bonvicini F, Manaresi E, Gallinella G, Gentilomi GA, Musiani M, Zerbini M Diagnosis of fetal parvovirus B19 infection: value of virological assays in fetal specimens. *BJOG*. 2009, 116, 6, 813-817

[55] Weir E Parvovirus B19 infection: fifth disease and more. *Canadian Medical Association J.* 2005, 172: 743

[56] el-Sayed Zaki M, Goda H. Relevance of parvovirus B19, herpes simplex virus 2, and cytomegalovirus virologic markers in maternal serum for diagnosis of unexplained recurrent abortions. *Arch. Pathol. Lab. Med.* 2007; 131(6):956-60.

[57] Modarows S and Dorsch S. Antibody responses in parvovirus B19 infected patients. *Pathol. Biol.* 2002, 50:326-31.

[58] Heegaard ED, Rasken CJ and Christensen J. Detection of parvovirus B19 NS1 specific antibodies by ELISA & Westren blotting employing recombinant NS1 protein as antigen. *J. Med. Virol.* 2002, 67:375-83.

[59] Söderlund M, Brown CS and Spaan WJ. Epitopes type-specific IgG response to capsid proteins VP1 & VP2 of human PB19. *J. Infect. Dis.* 1995:172:1431-36.

[60] Franssila R, Söderlund M, Brown CS, et al. IgG subclass response to human PB19 infection. *Clin. Diagn. Virol.* 1996, 6:41-49.

[61] Jeanne AJ. Comparison of a Baculovirus-Based VP2 Enzyme Immunoassay (EIA) to an Escherichia coli-Based VP1 EIA for Detection of Human Parvovirus B19 Immunoglobulin M and Immunoglobulin G in Sera of Pregnant Women. *J. of Clin. Microbio.* 2000, 38:1472-75.

INDEX

D

E

F

G

H

J

K

L

M

N

O

P

R

S

T